The Renewal of Nuclear Power in Finland

Energy, Climate and the Environment Series

Series Editor: David Elliott, Professor of Technology, Open University, UK

Titles include:

David Elliott (*editor*)
NUCLEAR OR NOT?
Does Nuclear Power Have a Place in a Sustainable Future?

David Elliott (*editor*)
SUSTAINABLE ENERGY
Opportunities and Limitation

Horace Herring and Steve Sorrell (*editors*)
ENERGY EFFICIENCY AND SUSTAINABLE CONSUMPTION
The Rebound Effect

Matti Kojo and Tapio Litmanen (*editors*)
THE RENEWAL OF NUCLEAR POWER IN FINLAND

Catherine Mitchell
THE POLITICAL ECONOMY OF SUSTAINABLE ENERGY

Ivan Scrase and Gordon MacKerron (*editors*)
ENERGY FOR THE FUTURE
A New Agenda

Gill Seyfang
SUSTAINABLE CONSUMPTION, COMMUNITY ACTION AND THE NEW ECONOMICS
Seeds of Change

Joseph Szarka
WIND POWER IN EUROPE
Politics, Business and Society

Energy, Climate and the Environment
Series Standing Order ISBN 978–0–230–00800–7 (hb)
978–0–230–22150–5 (pb)

The Renewal of Nuclear Power in Finland

Edited by

Matti Kojo
Researcher, Department of Political Science and International Relations
University of Tampere, Finland

and

Tapio Litmanen
Professor, Department of Social Sciences and Philosophy
University of Jyväskylä, Finland

palgrave
macmillan

First published 2009 by
PALGRAVE MACMILLAN

Palgrave Macmillan in the UK is an imprint of Macmillan Publishers Limited, registered in England, company number 785998, of Houndmills, Basingstoke, Hampshire RG21 6XS.

Palgrave Macmillan in the US is a division of St Martin's Press LLC, 175 Fifth Avenue, New York, NY 10010.

Palgrave Macmillan is the global academic imprint of the above companies and has companies and representatives throughout the world.

Palgrave® and Macmillan® are registered trademarks in the United States, the United Kingdom, Europe and other countries.

ISBN-13: 978–0–230–57592–9 hardback
ISBN-10: 0–230–57592–7 hardback

This book is printed on paper suitable for recycling and made from fully managed and sustained forest sources. Logging, pulping and manufacturing processes are expected to conform to the environmental regulations of the country of origin.

A catalogue record for this book is available from the British Library.

Library of Congress Cataloging-in-Publication Data

The renewal of nuclear power in Finland / edited by Matti Kojo and Tapio Litmanen.
p. cm. — (Energy, climate and the environment series)
Includes bibliographical references and index.
ISBN-13: 978–0–230–57592–9 (hardback)
ISBN-10: 0–230–57592–7 (hardback)
1. Nuclear industry—Finland. 2. Nuclear industry—Government policy—Finland. 3. Nuclear industry—Environmental aspects—Finland. 4. Nuclear power plants—Finland. I. Kojo, Matti, 1971– II. Litmanen, Tapio, 1963–
HD9698.F52R46 2009
333.792'4094897—dc22

2008050870

10 9 8 7 6 5 4 3 2 1
18 17 16 15 14 13 12 11 10 09

Printed and bound in Great Britain by
CPI Antony Rowe, Chippenham and Eastbourne

Contents

List of Figures and Tables

Figures

Tables

Series Editor's Preface

Concerns about the potential environmental, social and economic impacts of climate change have led to a major international debate over what could and should be done to reduce emissions of greenhouse gases, which are claimed to be the main cause. There is still a scientific debate over the likely scale of climate change, and the complex interactions between human activities and climate systems, but, in the words of no less than the Governor of California, Arnold Schwarzenegger, 'I say the debate is over. We know the science, we see the threat, and the time for action is now.'

Whatever we now do, there will have to be a lot of social and economic adaptation to climate change – preparing for increased flooding and other climate-related problems. However, the more fundamental response is to try to reduce or avoid the human activities that are seen as causing climate change. That means, primarily, trying to reduce or eliminate emission of greenhouse gases from the combustion of fossil fuels in vehicles and power stations. Given that around 80 per cent of the energy used in the world at present comes from these sources, this will be a major technological economic and political undertaking. It will involve reducing demand for energy (via lifestyle choice changes), producing and using whatever energy we still need more efficiently (getting more from less), and supplying the reduced amount of energy from non-fossil sources (basically switching over to renewables and/or nuclear power).

Each of these options opens up a range of social, economic and environmental issues. Industrial society and modern consumer cultures have been based on the ever-expanding use of fossil fuels, so the changes required will inevitably be challenging. Perhaps equally inevitable are disagreements and conflicts over the merits and demerits of the various options and in relation to strategies and policies for pursuing them. These conflicts and associated debates sometimes concern technical issues, but there are usually also underlying political and ideological commitments and agendas which shape, or at least colour, the ostensibly technical debates. In particular, at times, technical assertions can be used to buttress specific policy frameworks in ways which subsequently prove to be flawed.

The aim of this series is to provide texts which lay out the technical, environmental and political issues relating to the various proposed policies for responding to climate change. The focus is not primarily on the science of climate change, or on the technological detail, although there will be accounts of the state of the art, to aid assessment of the viability of the various options. However, the main focus is the policy conflicts over which strategy to pursue. The series adopts a critical approach and attempts to identify flaws in emerging policies, propositions and assertions. In particular, it seeks to illuminate counter-intuitive assessments, conclusions and new perspectives. The aim is not simply to map the debates, but to explore their structure, their underlying assumptions and their limitations. Texts are incisive and authoritative sources of critical analysis and commentary, indicating clearly the divergent views that have emerged and also identifying the shortcomings of these views. However, the books do not simply provide an overview, they also offer policy prescriptions.

One of the areas where there has perhaps been most conflict over policy choices in the energy field is in relation to nuclear power. Following the Chernobyl disaster in 1986, for much of Western Europe nuclear power was seen as an inappropriate option for the future, while renewable energy was seen as a better choice.

However, in Eastern and Central Europe this choice was seen as being much harder to make. Many of the ex-Soviet countries had large nuclear plants, and very little awareness of their renewable potentials. So although the EU required some of the old nuclear plants to be closed, as a condition of EU admission, this presented major longer-term problems – including the likelihood of expanded imports of gas and oil from Russia, as their newly liberated economies flourished. This issue came to a head particularly in countries like Lithuania, where the relations with Russia were strained, but it is also a more general problem. The International Energy Agency has argued that Central and Eastern Europe should become less dependent on Russian natural gas, and in some countries this issue is very significant. For example, 95 per cent of Slovakia's trade deficit results from paying Russia for the energy used by the country. This is the primary reason why Slovakia wants to go ahead with building the two new nuclear reactors to replace those shut as part of the EU accession process.

Meanwhile, with increasing concerns about climate change, and also growing fears about energy security, some Western European governments have begun to reconsider their attitudes to nuclear power. France, of course, had remained pro-nuclear, and is now building a new plant, and although most of the others, including Germany, Austria, Spain, Demark, the Netherlands, have remained anti-nuclear, the UK government has switched to a pro-nuclear stance and some others may follow, including Italy.

In this situation the decision by Finland to expand its nuclear programme has proved to be very influential. Viewed from the outside, Finland seemed to have ample renewable energy resources, and it is indeed one the leaders within the EU, obtaining nearly 30 per cent of its energy from renewable sources (compared, for example, to just over 2 per cent in the UK). Presumably, some observers felt, the argument was more strategic, akin to that made by the ex-Soviet countries: after all, Finland imports energy from Russia and might see that as a growing problem. Certainly many of the western EU countries were beginning to be concerned about the prospects of having to rely increasingly on Russian gas imports. Outsiders were also aware that Finland was undergoing significant economic expansion, and so its energy policy was clearly of wider interest to other EU countries trying to combine economic growth with climate-sensitive environmental sustainability.

This book provides some insights into how Finland reached its decision on new nuclear power plants. It looks in detail at the decision-making processes, rather than the technical issues as such or the wider geopolitical issues. The focus on the decision-making process highlights the way in which the conflicting views of various parties and lobbies interacted. It is interesting, for example, that the anti-nuclear movement, at one time quite strong, was unable to resist the pro-nuclear trend. It remains to be seen whether or not this pattern will be repeated in other western EU countries. Equally unclear is whether the Finnish programme will succeed as planned. The new reactor construction programme is already both years behind schedule and over-budget. Some critics see its economics as dubious – they argue that costs have been artificially reduced by the vendors and finance provided at low rates, in order to facilitate the construction of the EPR systems.

For a range of reasons, therefore, the Finnish programme is of wide interest, and this book provides if not all the answers to *why* it emerged, then certainly a detailed insight into *how* the decision emerged, with clear implications for decision making about contentious technological choices, democratic or otherwise, in other fields.

David Elliott

Preface

The book project was launched nearly four years ago. Both editors of this book had received many inquiries from journalists and fellow scholars from abroad. Interest in Finnish nuclear decisions in 2001 (nuclear waste) and 2002 (nuclear power) was widespread. A number of questions were raised concerning the decision-making process, risk evaluation, public participation, lobbying and transparency in relation to the nuclear decisions of the Parliament of Finland. Our problem was how to describe briefly, but analytically what were the crucial factors behind the recent nuclear power decisions. During our long and fruitful cooperation in the field of nuclear waste management research, we had pondered these questions on a number of occasions. Some of the answers had already been documented in our previous publications, but this time we thought that we needed to publish a book about nuclear power policy in Finland for an international community that was interested in the recent developments. This meant that a number of questions needed to be addressed, not only to respond to the information needs of the international media and policy makers, but also to reflect on Finnish society and its capacity for societal decision making.

The first step was taken in Summer 2005 as the editors proposed a workshop focused on nuclear energy policy for the organizing committee of the annual YHYS colloquium of the Finnish Society for Environmental Social Science. When the workshop proposal was accepted by the organizers, the editors started to invite researchers and PhD students studying the issue to present a paper. Although of societal debate and importance of the nuclear energy policy it became clear that there were few researchers and PhD students dealing with issue in Finland. The last large social scientific research programme focused on the nuclear energy issue and energy policy in general were conducted in the late 1980s. Thus, the focus group of the workshop was small and the interest to participate in was based upon participants' individual academic ambitions.

The second step, the annual colloquium of the Finnish Society for Environmental Social Science entitled 'Issues in Green Democracy', was taken in November 2005 in Turku. The papers presented in this seminar formed the basis for the first version of the manuscript. The third step, editing the manuscript, was a long one. For a variety of personal reasons it took well over a year before, in May 2007, the contributors held a small workshop at the University of Tampere to discuss the outline of the forthcoming book. The last and final step was taken in collaboration with Palgrave Macmillan.

Since the early days of the book project a number of Masters theses of social sciences on Finnish nuclear energy issues have been written. Hopefully, these works will produce a new and enthusiastic generation of Finnish researchers interested in studying critically different societal aspects of nuclear energy. For instance, at the moment we are witnessing how three energy companies are persuading both the Finnish decision makers and the public about the need of several new nuclear power plants. The possible expansion of the Finnish nuclear energy programme is an issue which needs to be discussed properly in the light of the long-term direct and indirect impacts on the development of the society. Furthermore, one of the companies is looking for an entirely new site for the nuclear power plant. The experiences of the existing nuclear communities evidence that the siting decision would have remarkable social impacts on the local community.

The editors of this book are grateful to the research programme 'The Power and Society in Finland' run by the Academy of Finland. The research funding granted to the joint research project of University of Tampere, Department of Political Science and International Relations and University of Jyväskylä, Department of Social Sciences and Philosophy by the Academy of Finland made it possible to finalize the book. The editors are grateful to the support offered by Palgrave Macmillan Associate Director and Head of Social Sciences Alexandra C. Webster and Senior Editorial Assistant Gemma d'Arcy Hughes of Palgrave Macmillan, as well as Editorial Services Consultant Nick Brock for his invaluable editorial assistance. Professor David Elliott at the Open University (UK) and the Series Editor of 'Energy, Climate and the Environment' helped us very kindly with the publishing process. The editors are very satisfied that the publisher was interested in providing international readers with a whole book on Finnish nuclear energy policy.

Over the course of many years the editors have had a privileged position to debate different energy and environmental policy issues with several scholars and students. We express our gratitude to all of them. In particular, we want to thank our nearest colleagues from Universities of Tampere and Jyväskylä: Emeritus Professor Olavi Borg, Professor Ilkka Ruostetsaari, Senior Research Fellow Pekka Hokkanen, Professor Esa Konttinen, Professor Marja Järvelä, Research Fellow Marja Ylönen and the late Professor Kaj Ilmonen.

Finally, the editors are most grateful to all of those who have contributed to the book. Without their commitment to this long project the book would not have been published.

Matti Kojo and Tapio Litmanen
Lapinlahti and Jyväskylä, February 2009

Notes on Contributors

Annukka Berg is a researcher and PhD student at the University of Helsinki, Department of Social Policy. Her Masters thesis dealt with discourses and actors of the 2000–02 parliamentary decision-making process in relation to Finland's fifth nuclear power plant. Berg's current research interests include deliberative policy analysis, environmental policy evaluation and politics on sustainable consumption and production.

Matti Kojo is a researcher and PhD student in Political Science at the University of Tampere, Department of Political Science and International Relations. Kojo has been involved with social scientific research on the Finnish nuclear waste management since 1997. He is editor and co-editor of three books dealing with nuclear waste and nuclear energy policies in Finland. His current research examines compensation negotiations as a part of the site selection process in Finland.

Harri Lammi is a PhD student in Environmental Policy at the University of Tampere and is writing his doctoral thesis about nuclear reactor decision making in Finland. Since 1996 Lammi has been following international and Finnish climate and energy policy as an university researcher and lecturer and a NGO energy campaigner. He followed the decision-making process on the fifth reactor in the Finnish Parliament in 2000–02 as an NGO campaigner. He is currently working as the programme director of Greenpeace Nordic.

Ari Lampinen is a nuclear physicist by training. He has worked since 1997 as an environmental scientist specializing in multidisciplinary energy and climate issues in universities, the private sector and NGOs, including nuclear energy technology and economics. He is currently an energy and climate consultant and fellow at Strömstad Academy in Sweden. He was involved in the fifth Finnish nuclear reactor issue both in the parliamentary review process as a representative of the University of Jyväskylä, and the Ministry of Trade and Industry review process representing Technology for Life, a sustainable development NGO of Finnish engineers and scientists.

Tapio Litmanen is a professor in Sociology at the University of Jyväskylä. His interests are in the areas of environmental sociology, the sociology of risk, and social movements studies. He has conducted several research projects in relation to risk and technology issues in Finland. For instance, he has published widely on nuclear power and nuclear waste issues. His doctoral thesis, 'The Struggle over Risk: The Spatial, Temporal, and Cultural Dimensions of

Protest Against Nuclear Technology', dealt with nuclear power disputes and anti-nuclear movements. He gives courses on environmental, risk and technology issues and supervises Masters and doctoral theses on issues related to his interest areas.

Erika Säynässalo is a researcher in Political Science at the University of Tampere's Department of Political Science and International Relations. She took her first steps as a researcher on the Finnish nuclear energy policy in the early 1990s after having become involved in the research project exploring the decision making of the fifth nuclear power plant from a local perspective. Her current research deals with Finnish nuclear energy policy from a comparative and multidisciplinary perspective. She has given numerous courses at her department on topics such as political institutions, comparative Nordic politics and Finnish nuclear energy policy.

List of Abbreviations

AREVA NP	French–German nuclear power company (earlier names Framatome NP and Nuclear Power International), www.framatome-anp.com
BWR	boiling light water reactor (the second most common nuclear reactor type globally)
CDM	Clean Development Mechanism of the Kyoto Protocol (article 12) of the UNFCCC
CEA	Commissariat à l'Energie Atomique
CF	Capacity Factor, the ratio of average annual power compared to maximal power
CHP	Combined Heat and Power production = cogeneration (all biopower in Finland uses this technology)
CLRTAP	UNECE Convention on Long-Range Transboundary Air Pollution (Geneva Convention of 1979 and its protocols)
DiP	Decision-in-Principle (in accordance with the Nuclear Energy Act)
EDF	Electricitè de France
EIA	Environmental Impact Assessment
ElFi	Lobbying and electricity market company of large Finnish electricity consumers, www.elfi.fi/en
EMV	Energy Market Authority (Energiamarkkinavirasto), www.energiamarkkinavirasto.fi
EPR	European Pressurized Water Reactor, a French–German PWR reactor type
ETS	emission trading scheme of the European Union
EU	European Union
Fingrid	Operator of the Finnish transmission grid, www.fingrid.fi/portal/in_english
Fortum	Energy company, the State of Finland is the biggest shareholder with a 50.83 per cent holding, the largest power producer in Finland with a 40 percent share, www.fortum.com
FPH	Fortum Power and Heat Ltd (formerly IVO), a part of the Fortum Consortium
GHG	greenhouse gas
GNP	Gross National Product
IAEA	International Atomic Energy Agency
IEA	International Energy Agency

IVO	Imatran Voima Ltd, 100 per cent state-owned power company established in 1932. Known as Fortum Power and Heat (FPH) since 1998
JF	Joint Fulfillment mechanism of the Kyoto Protocol (article 4) of the UNFCCC; it is used only by the EU based on the burden-sharing agreement
JI	Joint Implementation mechanism of the Kyoto Protocol (article 6) of the UNFCCC
Kyoto Protocol	The protocol of UNFCCC negotiated in Kyoto in 1997
LWR	light water reactor, reactor using enriched ^{235}U fuel
MP	Member of Parliament
Mt	million tons
MTI	The Ministry of Trade and Industry (Kauppa- ja teollisuusministeriö), from January 2008 the Ministry of Employment and the Economy (Työ- ja elinkeinoministeriö), www.tem.fi
MTK	the Central Union of Agricultural Producers and Forest Owners (Maa- ja metsätaloustuottajain Keskusliitto)
MW_e	Megawatt electric, electric power output
MW_{th}	Megawatt thermal, total power output (including waste heat)
NGO	non-govermental organization
NPP	nuclear power plant (may include one or several nuclear reactors)
OECD	Organisation for Economic Co-operation and Development
Posiva	Nuclear waste company, established in 1995, owned by TVO (60 per cent) and Fortum Power and Heat (40 per cent) www.posiva.fi/englanti
PVO	Power company (Pohjolan Voima Oy) that owns the majority of TVO shares, the forest industry companies UPM Kymmene and Stora Enso are the biggest shareholders of PVO, www.pvo.fi/en-GB/home/
PWR	pressurized light water reactor (most common nuclear reactor type globally)
SAK	the Central Organization of Finnish Trade Unions (Suomen Ammattiliittojen Keskusjärjestö)
SNF	spent nuclear fuel
STUK	Radiation and Nuclear Safety Authority (Säteilyturvakeskus), www.stuk.fi/en_GB/
TEKES	Finnish National Technology Agency, administered by the MTI, www.tekes.fi

TPES	total primary energy supply
TT	the Confederation of Finnish Industry and Employers (Teollisuuden ja Työnantajain Keskusliitto), nowadays a part of the Confederation of Finnish Industries (EK)
TVO	Condensing nuclear and coal electricity generation company (Teollisuuden Voima Oyj), www.tvo.fi/index_eng.shtml
TWh_e	TWh electric, terawatt-hours of energy as electricity
VTT	Technical Research Centre of Finland, www.vtt.fi/index.jsp?lang=en
UN	United Nations
UNECE	United Nations Economic Commission for Europe
UNFCCC	United Nations Framework Convention on Climate Change: the UN climate convention negotiated in the UN summit in Rio de Janeiro; it came into force in 1994
VVER	Russian PWR reactor type

Part I
Introduction

1

The Transillumination of Finnish Nuclear Policy: Seeking a Shortcut to a Low Carbon Society

Tapio Litmanen

At the beginning of the new millennium Finland's national energy policy, in common with other nation-states, was at a crossroads. The Finnish decision was to allow the construction of a new nuclear power plant (NPP), a development which can be regarded as unique in international terms. In May 2002 the Finnish Parliament ratified the government's earlier favourable Decision-in-Principle (DiP) on a fifth NPP unit.[1] In December 2003 the power company Teollisuuden Voima (TVO) made an investment decision about the European Pressurised Water Reactor (EPR), which has a net electrical output of about 1,600 MW. This plant is being built by a consortium of Framatome ANP and Siemens AG.[2] The country also has an international reputation as a pioneer of nuclear waste management. In May 1999 Posiva, the company responsible for the final disposal of spent nuclear fuel in Finland, suggested that the government of Finland should consider only Olkiluoto in Eurajoki in its application for a DiP as the final disposal site. In January 2000 the municipal council of Eurajoki made a positive statement on the DiP. The government made the DiP on 21 December 2000, and its decision was ratified by Parliament on 18 May 2001.

This nuclear power decision making was met with acclaim from various parties. For instance, the Chatham House Report states that 'The Scandinavian model of decision-making is characterized by a relatively consensual approach to cross-party politics and considerable public trust in government and regulators...' (Grimston 2005, 43–4).[3]

In recent years there has been increased international interest in Finnish energy policy. In addition, the country's energy policy has undergone a number of changes. The electricity markets have been liberalized very rapidly, with Finland being among the first countries to do so, even before the EU directive came into force. Today Finland is part of Nord Pool, the largest electricity market in Europe. During the 1990s Denmark, Finland, Norway and Sweden cooperated to form a unique multinational electricity market.

The national electricity markets were opened up to cross-border trade and a common power exchange. This reform began in Norway in 1991, continued in Sweden in 1996 and Finland in 1998, and was completed by Denmark in 2000 (Holttinen 2005; Amundsen and Bergman 2007). Furthermore, as a result of its membership of the European Union Finland can be viewed as an interesting case study for the examination of energy policy power and governance. In recent years the EU has increased its efforts to coordinate national energy policies; for instance, during Finland's EU Presidency in 2006 one of the main objectives for the work of the European Council was to develop the EU's energy policy.

Because of developments in Finland's energy sector representatives of its nuclear power industry, governmental authorities and politicians have been sought-after speakers at seminars throughout the world. Their task has been to analyse these unique decisions so that people can understand what is so special about the Finnish model of decision making on controversial energy policy issues. For instance, the former long-standing Prime Minister Paavo Lipponen has been active in promoting national energy policy and the country's very open and democratic decision-making processes.

> The Parliament made its decision on 24 May 2002 to regard the project to be 'in accordance with the overall good of society'. The government of Mr. Matti Vanhanen granted the construction licence in February 2005. The operating licence that is needed before the starting the commercial operation in 2010 ... 2011, will also be subject to government approval. The very democratic and open decision process, aiming at the broadest possible social and political consensus, has turned out to be a great strength to our energy policy. Consensus was enhanced by combining the nuclear decision with a programme of developing more domestic and renewable sources. The Economic Affairs Committee of our parliament deliberations were extremely thorough – the Committee heard dozens of experts, including representatives of environmental NGOs. (Lipponen 2007)

Metaphorically speaking, the floodgates were now open. The debate about a sixth nuclear reactor is ongoing and the private electricity industry is planning its own NPP, which could be the seventh. The three major Finnish political parties – the Centre Party of Finland,[4] the Finnish Social Democratic Party[5] and the National Coalition Party[6] – are currently formulating their stance in respect of these new plans.

This book offers a critical analysis of Finnish nuclear energy policy, which seeks to address some of the shortcomings of the one-dimensional picture of the country's nuclear power policy that has been advanced in much of the other literature. The understanding of nuclear power decisions needs more thorough analysis than that which appears in rapid policy evaluations or glib diplomatic speeches. We hope that the book will offer a more comprehensive description of how the acclaimed Finnish democracy functions and what other instruments of power are in use. Thus, the critical stance adopted in this book may also contribute to the development of policy studies. However, our aim is not to make a detailed and systematic study of policy arrangements in the energy policy field, but rather to highlight some recent features relating to how the governance of the field functions. We concur with those scholars who emphasize the complexity of policy governance. For us it is important that there should be a focus on the roles of actors, actor coalitions, resources, the rules of the game, policy discourses and socio-techno-cultural institutional circumstances.

Living through the turbulent years of the first decade of the new millennium we have witnessed how the importance of energy policy has grown and how the role of active energy policy has been re-evaluated in domestic, security and foreign policy fields. The end of the Cold War presented an opportunity for the ideology of market liberalism in the 1990s, but developments after the September 11 attacks have forced governments to take a more active role in energy policy. Today state control and the state ownership of crucial infrastructures and resources are seen in a very different light. It is difficult to give an exact description of today's dominant energy policy orientation, but in the context of this book suffice it to say that most of the actors involved have a considerable stake in the business. Our aim is not to analyze global energy policy developments, but to try to understand recent changes in Finnish nuclear policy in the context of domestic and international energy policy trends.

The difficulty of establishing a coherent picture of energy policy

Our understanding of energy policy goes beyond the conventional definition. It is rather narrow to regard energy policy as a plan of action for tackling issues related to energy supply, demand, the development of energy-related industries but also the different environmental and social implications of energy activities. It may be suggested that it is inappropriate to talk of energy policy in the singular. Even within a single nation-state there are a number of influential actors, such as governmental authorities, institutions or more locally based actors in addition to the energy industry, which cannot be seen as monolithic actors. Each of these actors may exercise their own policies or policies, which may differ somewhat from the measures of other actors. Policy measures taken by these different entities may not be in concert, and the aims of the actors may vary – some of them can be acting in the benefit of public interest and serving the general interest of the society, whereas others serve narrower interests. A complex combination of different measures, aims and interests at a certain point in time may justify us in claiming that there is something we can call national energy policy. We can, of course, draw a still-life picture from those conventional national actions, such as the legislation on commercial energy activities (trading, transport, storage, and so on), the legislation governing energy use (for example, efficiency standards, emission standards), instructions for state-owned energy sector assets and organizations, and fiscal policies related to energy products and services (taxes, exemptions, subsidies). The real problem in any discussion of national energy policy is how to include this complex configuration of different actors pulling in a variety of directions into one single picture. In addition, the fluidity of policies is evident as important policy actors make decisions and take actions all the time. To construct a comprehensive understanding of the issues on which the actors have reached consensus is easier than in those cases where a lack of congruity or absence of coherence is evident. For instance, the Climate Strategy for 2001 was under reconstruction in Autumn 2005 in order to take into account the emissions trading in the European Union and the Kyoto mechanisms, but because of continual dissension in various ministries and political parties on energy policy questions its completion was delayed by months. From this example we can observe that national policy is also contingent upon international policy measures such as international environmental treaties and agreements. The important factors include not only international energy sector treaties and alliances, but also general supranational trade agreements and intergovernmental legislation such as the key European legislation to establish the Internal Market of Electricity (The Electricity Directive 2003/55/EC of the European Union).

One way to understand energy policy in modern societies is to consider the policy arrangements. For instance, van Rooijen and van Wees (2006) apply this concept when trying to understand policy changes in the field of green electricity (see also van Tatenhove, Arts and Leroy 2000). They use it to refer to the way in which a policy field is organized in terms of content (the policy discourse and programmes) and organization (actors, coalitions, power, influence and rules). For them actors are organizations or institutions that operate in a specific policy domain, such as the promotion of green electricity. They use the term 'coalition' to refer to a group of actors who share broadly similar policy goals and programmes and whose common objectives form the basis for their involvement in the policy process. They perceive that the determinants of power and influence are actors' relationships, particularly their interdependency and the distribution of resources among the actors. Resources can take the form of financial means, knowledge or media access, and their distribution indicates the relative influence of actors at various stages in the policy process. The policy arrangements also include formal and informal rules, which provide a framework for policies and political actions and determine which norms are regarded as legitimate. Formal rules are much easier to comprehend, precisely because they are formally agreed to; however, informal rules reflect the dominant political culture and it is therefore more difficult to trace them empirically. A shared understanding of single events or certain choices requires an established policy discourse, which is a specific combination of ideas, concepts and categorizations produced, reproduced and transformed in a particular set of practices and through which meaning is given to physical and social realities. Van Rooijen and van Wees (2006) state that a policy discourse encompasses the content of policy in terms of norms, values and the specific content of policy papers and measures.

Arts and van Tatenhove (2004) describe how the vocabulary of policy and policy studies has undergone a process of renewal. Making reference to several scholars, they point out how certain terms, such as governance, networks, institutional capacity, discourse coalition and deliberation, have gained ascendancy in scientific language over earlier terms, through which we perceive the world from the perspective of concrete actors or isolated power centres. Their suggestion is that an enthusiasm to adopt new vocabulary should not cause us to overlook important concepts, such as power. Their theoretical aim is to strengthen the analytical power of their own conceptual innovation, the policy arrangement approach, by combining it with the old power perspective. As they state, their programmatic task has been the creation of a policy arrangement approach to understand recent shifts in governance and power. The basic assumptions of this approach are: (1) the institutional embeddedness of multi-actor policy processes; (2) the manifestation of structural developments, such as globalization, in concrete policy practices; (3) the role of different faces of power in policy making; and (4) the importance of both substance and organization, as well as of change and continuity in policy practices. They are interested in the shaping of the policy domain. Therefore they stress the importance of substance (such as principles, objectives and measures) and organization (such as departments, instruments, procedures and the division of tasks and competence).

This approach also subsumes the idea of dynamic processes, as they explain how any shape of a policy domain is only temporary, because arrangements are constantly under pressure to change. The other dimension besides time is space. Arrangements are not universal, but are characterized by specific spatial boundaries. They can be found at different levels of policy making – local, national and transnational – or interconnect these levels, e.g. as specific forms of multi-level governance. The substantial and organizational characteristics of a policy arrangement can be analysed on the basis or four dimensions: policy coalitions, 'rules of the game', policy discourses, and resources. The number of players who share resources and/or interpretations of policy discourse in the context of the rules of the game can be regarded as policy coalitions. These coalitions identify more or less similar policy goals, and engage in policy processes to achieve them. The authors distinguish between supportive and challenging coalitions. The rules of the game refer to legitimate norms and the political weight of issues, agendas, interests, policies, decisions and measures. They can be characterized as procedure guidelines giving instructions from the right political discourse and right players for a proper constitution of the field. The third dimension of policy arrangement is policy discourse, which is defined as a dominant interpretative scheme, ranging from formal policy discourse to popular story lines by which meaning is given to a policy domain. The final dimension is resources related to the concept of power, meaning that more or less permanent capacity of agents-in-interaction to achieve certain policy outcomes, but also dispositional and structural phenomena or a social and political system. The authors share the view that power is also concerned with the asymmetrical distribution of resources in society (domination as a structural phenomenon), revealing itself in positions of autonomy and dependency between actors (dispositional phenomenon) (Arts and van Tatenhove 2004, 339–43).

The history of Finnish energy policy

Since the earliest days of Finnish energy policy the issue has been regarded not only from the point of view of economic and technological investments but also as an important factor in the modernization of Finnish society. Studying the electrification of Finland from 1877 to 1977 Myllyntaus (1991, 284) describe how rapidly the country's electricity consumption increased.

> ... Finland leapfrogged over some phases of development, direct from an archaic subsistence economy to hectic industrialization. In Chapter 4, it was claimed that Finland reached approximately the same level of electricity consumption per capita in 1900 as Britain or some major Continental countries, such as France and Italy. After some setbacks between 1900 and 1925, Finland rose to be among the top ten electricity consumers in per capita terms during both the interwar years and postwar period. This performance was fairly exceptional, because of the late industrializing countries only Iceland, New Zealand and Finland managed to enter the group of the top ten electricity users (in per capita terms) between 1925 and 1975.

Myllyntaus emphasizes that the relationship between rapid industrialization and electrification was a mutually supportive one and that there was a societal consensus to modernize the country rapidly. According to him (Myllyntaus 1991, 285–8), the factors behind this atypical brisk development were: (1) general cultural and societal factors; (2) cultural resources; (3) extensive use of indigenous energy sources other than coal and oil; (4) early integration into the world market; and (5) the energy-intensive structure of industry. Over the years public opinion in relation to industry and new technology has been positive. For instance, Finnish peasants adopted the capitalist way of thinking at a very early in the development of society and entrepreneurship and the introduction of innovations were seen as virtues – or even as a national duty. Thus a favourable attitude towards industrialism has been part of Finnish nationalism for many generations. Consensual thinking about the matter among a variety of interest groups has accelerated this development. Cultural resources, such as a well-developed education system, an active media and networks to educate experts abroad before Finland's own expertise expanded were necessary factors in the electrification of Finland. Although the country lacked reserves of coal and oil, it had massive indigenous resources of timber and peat. The processing of these resources led to the adoption of new technology, but it was hydropower which became the main indigenous energy source for electricity generation at the turn of the nineteenth century. Its early integration into the world market hastened the development of energy technology. The export of sawn timber, pulp and paper strengthened the Finnish economy and advanced the overall process of industrialization. New machinery and technology were needed as well as energy to run the economy. The fifth factor to which Myllyntaus refers is related to the previous one, because the Finnish wood processing industry is extremely energy-intensive. In summary, Myllyntaus argues that these factors contributed to the rapid early development of electricity technology and business.

In the earliest energy policy formulation the main interest was in increasing electricity production through the use of hydroelectric power (Myllyntaus 1991, 130). In point of fact there were two separate energy policies: electricity policy and fuel policy. The fuel economy was principally seen as a separate issue of foreign trade, indigenous fuels and employment. Electricity and domestic heating were considered to be separate issues. As recently as the 1970s a split approach to energy led to a crisis. The oil crises of 1973–74 and the growing share of thermal power compelled decision-makers to formulate an integrated energy policy. The main objectives of electricity policy were low price and reliability, and imported coal and oil largely replaced indigenous fuels. The oil crisis led to a reformulation of the objectives: the security of energy supply, the conservation of energy, an increasing share of indigenous energy sources, management efficiency and the state planning of the national energy economy (see also Nurmi 1980).

During the 1970s and 1980s the most important principle in Finnish energy policy was regulation and central planning. Planning ideology was strong until the 1980s, which saw the emergence of more liberalistic trends in the energy sector. Ruostetsaari (1998) describes how this decade was a period of two competing ideologies. Gradually state-centred planning and regulation ideology had to make more room for market mechanisms and open competition. During the neo-liberal phase of the 1990s the dominant principle was deregulation, that is, a dismantling of the energy sector's regulation and its replacement by free markets. Both Ruostetsaari (1998) and Pineau and Hämäläinen (2000) have documented this historic change in energy policy. They all agree that the Electricity Market Act (EMA), passed in 1995, forms a milestone in increasing competition and decreasing regulation in the Finnish electricity industry. The objectives of EMA were to increase the levels of efficiency and competition in power generation and transmission in order to prepare the Finnish electricity market for opening up to international competition. Decision makers also anticipated new EU energy policy developments and wanted the country to participate in the creation of a Nordic electricity market. According to Pineau and Hämäläinen (2000, 186), EMA had several consequences. In terms of this book perhaps the most important ones are the complete opening of the market, the gradual opening of networks and the creation of Fingrid.[7] Regarding the overall conclusion of the reform the authors point out that even before the reform the market in Finland was liberalized and that one should not hesitate to make simplifying inferences.[8] They state that a comparison of the Finnish electricity market before and after the restructuring revealed that some vertical breakdown had occurred – for example, the creation of Fingrid – but at the level of power generation the horizontal integration maintained the same, as did the ownership. The coordination of sales became greater following the opening of the unique Nord Pool spot market. System operations and transmission came under the sole control of Fingrid. At the distribution level, mergers even increased the level of horizontal integration, but market type and ownership remained the same. Rather surprisingly, it was found that the so-called regulated market contained several uncommon features. The earlier emphasis on central planning and regulation did not prevent a considerable degree of diversity in generation and competition, in terms of technologies and the number of producers. In addition, the transmission field was not a monopoly and even without regulatory agency supervision[9] the actors behaved reasonably and engaged in cooperation.

Ruostetsaari (1998, 190) stresses that in the 1990s the political parties also internalized and accepted the principle of competition.[10] One illustration of this was that during the parliamentary processing of the new electricity markets law (1995) and during the discussion on the fusion of two large state-owned energy companies, IVO and Neste, in 1998 the politicians did not bother to make speeches or get involved in the preparation. Ruostetsaari (1998, 200–2) reports that politicians were not interested in these big energy policy questions, even though these decisions led Finnish energy policy towards the adoption of a market ideology. The result of political apathy in respect of energy policy questions meant that the role of government officials had been strengthened. Therefore the principles of energy policy were shaped in the ministries when separate decisions were prepared by key officials, which overcame the political void. Kyllönen (2004), Lammi (1994) and Kojo (2004) give a description of how during the preparation of the governmental DiP for a new nuclear reactor the Ministry of Trade and Industry (MTI) and certain officials acted in a particularly goal-oriented manner. In addition to the preparation of the nuclear power decision the ministry was in charge of the formulation of climate strategy. They purposefully constructed two scenarios in which the export of natural gas was an alternative to nuclear power in preventing the increasing production of greenhouse gases. Environmental organizations demanded that the ministry announce the assumptions of their calculative scenarios, but the head of the energy office refused to do so. The authors criticize the behaviour of both civil servants and leading experts in the preparation of nuclear power and climate strategy decisions. Their conclusion is that the openness and transparency of decision-making was not what it should be in a democratic society.

Today's energy policy

This power shift in energy policy issues did not mean, however, that the political parties abandoned all interest in energy policy. Ruostetsaari (1998, 212–15) recognizes that since the 1980s Finnish energy policy has been characterized by three strong party-related subsystems: domestic energy sources, natural gas and nuclear power.[11] The first and most obvious political coalition is around the issue of domestic energy sources and, in particular, peat. The Centre Party of Finland, the Finnish Rural Party,[12] the Finnish Christian League[13] and partly also the Swedish People's Party[14] have shared the aim that state should promote and subsidize both peat and wood as energy sources. The Green League[15] is something of an outside member of this subsystem, because they favour the use of wood as an energy source, but are concerned about the use of peat as an energy source because of the possible environmental damage. The other ideology of this coalition has been their anti-nuclear stance, which is in opposition to the nuclear power subsystem. The Finnish Social Democratic Party and particularly the National Coalition Party have been keen supporters of nuclear power. Their lack of interest in domestic energy sources is based on the fact that peat and wood are produced in rural areas, which is the power base of the Centre Party of Finland. These three parties usually compete for the position of the biggest party in Finland. The coalition around nuclear power is perhaps the strongest, because in addition to these influential parties it also includes power companies, the majority of the labour unions and almost the whole of industry and commerce, along with their representative organizations. The role of the third energy policy subsystem around natural gas has disintegrated gradually. Traditionally the support for this energy source has increased from political right to left, but the collapse of the Soviet Union changed the situation. During the previous nuclear debates natural gas exported from Russia had been seen as an alternative to the further construction of nuclear power facilities. This argument about increasing the exports of natural gas has also been supported by the political parties of the domestic energy sources subsystem. The Left Alliance[16] and its predecessors have been particular advocates of this energy source. As a consequence of the restructuring of Finnish economic life during the

1990s the argument in favour of increasing energy exports from Russia and also increasing the level of energy dependence was not viewed with delight. Today the Left Alliance also speaks for domestic energy sources and supports the development of Finnish nuclear power.

The present-day objectives of energy policy are: security of supply; effective energy markets and economy; environmental questions and safety (MTI 2005; IAEA 2003, 280). The International Energy Agency (IEA),[17] which reviews national energy policies, has praised Finland's policy, stating that it is characterized by three commendable traits: holistic strategy, utilization of the international energy trade, and minor share of energy regulation (IEA 2004a). First, they state that a holistic approach to energy means that energy policy strives to pursue the three E's simultaneously – Energy security, Economic development and Environmental sustainability. The approach is perceived as an effective way to communicate on energy policy issues between the various ministries and energy programmes. The reviewers emphasize that in an ideal case these actors and programmes can pursue numerous policy goals in tandem and act in concert rather than in conflict with one another. Second, in pursuit of low energy costs and security of supply, Finland uses the international trade system. Because of a lack of substantial domestic energy resources the country imports fuels and electricity from a variety of countries. For example, Finland is part of the Nordic Power Market (Nord Pool), which means that it has common electricity markets with other Scandinavian countries. Third, national policy contains very little regulation. For instance, the IEA has found that its electricity sector is one of the least regulated in the world, with companies free to build power plants as they wish and all of the customers free to choose their own supplier. Although the agency is in favour of deregulation they recommend that the regulatory authority should also be expanded, e.g., in the field of high-level nuclear waste management and electricity transmission and distribution (IEA 2004a, 9–12). In addition, energy supply decisions on energy systems take place at a fairly decentralized level. A substantial proportion of energy is imported and produced by private enterprises and the state-owned energy companies are also run on a commercial basis (IAEA 2003, 280).

Recently in a press conference in Helsinki IEA praised Finland's commitment to balanced and realistic energy policy.[18] They also published a country review report (IEA 2008) 'Energy Policies of IEA Countries – Finland – 2007' which concluded that:

> A small, somewhat isolated country, Finland takes a balanced view of its energy policy, taking advantage of its situation wherever possible. [...] For example, it makes straightforward use of European Union directives and policies, without adding too many extra layers of regulations that could undermine the effectiveness of continent-wide policies. A small electricity market by itself, the country makes extensive use of gains from trade

> with its neighbours, not only as part of the well-functioning Nordic electricity market, but also with connections to Russia and the Baltics. [...] The greatest of these challenges may be energy supply security – its very high import dependence in general and its strong reliance on one import source in particular. To balance the lack of import diversity, the country has high domestic diversity, drawing its supply from many different sources, including domestic sources – namely nuclear, biomass and hydro. Finland, like most IEA countries, also faces the challenge of rising greenhouse gas emissions [...] Here, again, the country has taken a pragmatic approach. It has taken as much advantage of domestic and international trading as possible, allowing it to lower its own cost of compliance, as well as driving international development of a market for carbon emissions ... (IEA 2008)

In order to understand Finnish energy policy one must also pay attention to some basic facts of energy consumption, resources and supply. It is characteristic of the Finnish energy field that on the one hand the primary fuel supply is diverse and on the other the country's industrial structure is very energy intensive. Its relative dependence on foreign energy supplies can be seen from the fact that crude oil and oil products constitute a major part of imported energy. Other main fuels imported to Finland are coal and natural gas (IAEA 2003, 278). The IEA (2004a) report states that Finland's primary fuel supply is a sensible mixture of different energy sources. The largest contributor is oil, which in 2001 accounted for 28 per cent of total primary energy supply (TPES). Four other fuel sources contribute at least 10 per cent each, including biomass (20 per cent), nuclear power (18 per cent), coal (12 per cent) and natural gas (11 per cent). One peculiar feature of Finnish energy policy is that, although the share of renewable energy in Finland is relatively high as a result of the use of by-products and waste from the massive forest industry sector, the share of other renewable sources, such as wind power, has remained low by international comparison. The fuel mix is projected to become further diversified as the share of oil of TPES falls to 24 per cent by 2010 with gains made by nuclear, coal and natural gas. Wind power as a source of energy increased its output by 200 per cent between 1998 and 2001. Nevertheless, its contribution remains small, accounting for only 0.02 per cent of TPES in 2000. By 2010, it is expected to represent 0.08 per cent of TPES and by 2020, 0.14 per cent.

The other side of the coin is that the total primary energy consumption per capita was about 65 per cent higher than the European Union average (according to 2001 statistics) and about 39 per cent higher than the OECD average. This is mainly the result of Finland's cold climate, which demands that houses are heated throughout much of the year, and the structure of its industrial base, which is dominated by energy-intensive processing industries (wood, especially paper, heavy metal and chemicals). A third factor is the relatively high transportation requirements per capita caused by the low levels of population density (IAEA 2003, 277). The primary indigenous energy resources in Finland are hydroelectric power, wood, wood waste, pulping liqueurs and peat. Indigenous fuels and hydroelectric power cover about 30 per cent of the energy demand. Finland imports all of its oil, natural gas, coal and uranium. The percentage shares of primary energy sources in 2002 were: oil 26, coal 13, natural gas 11, indigenous fuels 27, hydro power 3, nuclear power 17, net electricity imports 3 (IAEA 2003, 278).

In 1994, Finland ratified the Framework Convention on Climatic Change. According to the 'Burden Sharing Agreement' among EU Member States, Finland has agreed to keep its greenhouse gas (GHG) emissions at 1990 levels during the 2008–12 Kyoto target period. For the reference year (1990) of the convention, the total CO_2 emissions in Finland were 53.9 million tons (Mt) and the total amount of greenhouse gases was 71.5 Mt. Meeting the emission limits seems to be a challenging task, because, for instance, between 2000 and 2002 CO_2 emissions increased by almost 15 per cent. According to Lampinen (2005), in 2003 the figure of total GHGs was 86 Mt, which meant that it was more than 20 per cent higher than the total in the reference year. Finland was among the poorest industrial nations in reducing greenhouse gases.[19] According to the IEA, the most important factors behind this volatility are the climate and energy-intensive industry. The climate increases the cost of heating, but can also reduce the amount of hydroelectric power available in Finland (20 per cent of the country's electricity) and, what is also crucial, the availability of inexpensive imported electricity from the other Nordic countries, primarily Norway. Another important factor is the number of energy-intensive industries competing in international markets, such as the pulp and paper industry and metals. The ebbs and flows of international markets affect the emissions so that increased levels of industrial production increase emissions – and vice versa. 'While these two factors – the weather and economic activity – affect the emissions paths of all countries, they are particularly important for Finland' (IEA 2004, 29).

Finnish environmental organizations and some scholars do not agree with the opinion of the IEA and the MTI that Finland is unable to reduce its CO_2 emissions. Lampinen (2000) has analysed those six arguments[20] supporting non-reduction policy and he finds that they do not withstand critical analysis. Although they are partly valid, Lampinen stresses that changing the assumptions behind the arguments may lead to different conclusions in which the reduction of emissions might produce benefits and open up opportunities. Kaivo-oja and Luukkainen's (2004, 358) analysis also indicates that some commonly used arguments may not be valid. The authors show that in recent decades there have been no significant improvements in the level of Finland's total energy efficiency. Comparing the amount of energy that was used to produce one Finnmark of GDP in 1960 with the amount required in 1998, they point out that there has been no change towards less energy-intensive production modes. On the contrary, the 1990s were a period of reindustrialization, which meant that energy efficiency did not improve. Finland is one of the EU Member States that has not improved their level of energy efficiency. To a large extent, France, Sweden and Finland have relied on nuclear power for their supply and on energy-intensive industrial development of their economy (Kaivo-oja and Luukkainen 2004, 380). In contrast to this the most important changes in Finnish CO_2 emissions happened in the 1960s, when a fuel switch towards carbon-intensive energy production took place and in the late 1970s and early 1980s, when the introduction of four NPPs caused remarkable fuel switch towards less CO_2-intensive energy production. Since then there has been non major fuel switch in Finland. One striking feature in their analysis is the substantial fluctuations in terms of energy and CO_2 intensity. They have been caused principally by changes in hydroelectric power production in the common Nordic market. As in Denmark, the domestic coal-based condensing power production adapts to changes in hydroelectric power supply and fluctuations.

In summary, Finnish industry accounts for a higher proportion of total energy consumption than is the case in other OECD countries. As Kara and Tuhkanen (2002, 14) explain, just over half of both primary energy and electricity is currently consumed by industry. In addition, the pulp and paper sector alone accounts for more than half of all industrial consumption. This is also the reason for the large share of wood biomass, about 20 per cent of primary energy consumption, since more than two-thirds of this consumption takes place in the pulp and paper mills. Kara also points out that another specific feature of the Finnish energy system is its high overall efficiency in energy production, since about one-third of its electricity is produced at the combined heat and power plants. They are either connected to the district heating systems of communities or supply process heat and steam to industrial installations.

Finnish energy policy in an international context

As we emphasize that energy policy is not a custom fixed by tradition, we must pay attention to different forces that are shaping – or trying to shape – national energy policy. Fairly interesting developments are taking place on national and international electricity markets: the reconsideration of the role of nuclear power in reducing atmospheric greenhouse gases, the opening up of the national electricity markets to international competition and attempts to create markets for alternative energy sources. Alongside other developments, these have had a considerable effect on energy policy debates in different countries. Reactions, responses and policy measures to these vary, but almost all actors in the field of energy policy have to consider their attitudes to these issues.

At the national level countries are pondering the effects of the Kyoto Climate agreement on the production of electricity and energy policy. As Morland (2001) suspected, the signing of the agreement has led to a reopening of the debate over the link between climate change and nuclear power. The prevailing assumption – that nuclear power was a source of energy which seemed to belong to the past and was not to be developed in contemporary societies – has been reconsidered (cf. Elliot 1997; Toke 2005; Blowers 2007). Many countries, including Sweden and Germany, had believed that nuclear power should be phased out with no thought being given to the energy sources that could replace it. At the same time the Member States of the EU have drawn up a common strategy on climate in which the most important policy instrument is the EU Emissions Trading Scheme (EU ETS), which was initiated in 2005. Under this scheme, enterprises can buy and sell emissions allowances, so that the level of emissions may increase in one country while at the same time decreasing in another. Governments are making calculations about how to meet the need for electricity without increasing CO_2 emissions, without spending too much on emission trading and without building new NPPs. As Morland (2001, 58) states, not even the most fervent proponents of nuclear power claim that a reliance on nuclear energy alone will help to achieve the necessary reductions in carbon dioxide gases. But the developments he was anticipating occurred in the UK in April 2005, when the advisors to Prime Minister Tony Blair suggested that the construction of new nuclear power stations would be the best way to meet the country's targets for reducing the emissions of gases responsible for global warming (Toke 2005, 55).

The recent history of the British nuclear programme is rather interesting, because as in Finland in 1991 so also in Britain in 1990 the nuclear power option was cancelled and re-opened within the course of just a few years. Toke (2005, 54–5) explains that the newly privatized electricity industry cancelled plans for the construction of three new NPPs despite the existence of a 'fossil fuel levy' intended to subsidize nuclear power, and despite the fact that, as measured by opinion surveys, anti-nuclear attitudes were decreasing. Toke argues that even for the Conservative government who had promised to review energy policy, it would be very difficult to fund new nuclear power through a privatized electricity system. A more detailed picture was drawn up by Winskel (2002), who provided an analysis of various phases of British electricity markets and nuclear policy, moving from the 1920s and ending up with a consideration of the developments of the 1990s. His conclusion is that between the mid-1950s and the mid-1980s nuclear power exerted a powerful influence on British energy policy, but that between 1987 and 1995 the process of privatizing the British Electricity Supply Industry (ESI) marginalized the role of nuclear power in Britain. With no subsidies, and no government funds provided for the construction of new nuclear plants and furthermore in a condition of uncertainty about long-term electricity prices, the still state-owned British Energy was unable to proceed with the construction of any new NPPs (Winskel 2002, 453–4). According to Winskel, private investors played a crucial role in bringing about the end of the British nuclear power programme. The return of a Labour government in the late 1990s suggested that there would be a bleak future for nuclear power, but the growing awareness of economic consequences, the Kyoto Protocol and the UK's own CO_2 emission targets led them to reconsider the nuclear option.

Many countries in both Europe and North America have started to integrate their electricity sectors into larger communities. For instance, Europe is currently in the process of creating common electricity markets. However, after five years of competition the EU has noted that many aspects of implementation are still disappointing. In October 2004, EU sent a letter warning 18 Member States that they had still failed to comply fully with the Commission in respect of the legal measures taken to transpose the latest Directives (EU 2005). This finding has been verified by van den Hoven and Forschauer (2004) whose study suggests that the liberalization of the electricity markets[21] in western countries has not succeeded because of strong national interests which have limited the level of domestic electricity market reforms. Their comparative analysis of France and Canada demonstrates that national interests may still prevail in an era of intergovernmental formulation of electricity policy. The EU report lists four reasons for the failure to create common electricity markets. First, foreign companies have been unable to penetrate national markets.[22] In most cases, foreign suppliers represent less than 20 per cent of the market share. Secondly, Member States are still failing to deal with the issue of market structure. The electricity markets in too many Member States are dominated by one or two companies, and there is often inadequate capacity for cross-border competition. Third, energy suppliers are still dominating the electricity distribution system. The EU argues that fully independent transmission system operators and regulators without connections to supply companies can ensure cost-reflective tariffs and the removal of any cross-subsidies. Fourth, one obstacle to the internal market is the continued existence of regulated end user prices for electricity and gas alongside the competitive market and associated long-term power purchase arrangements (PPAs).

Amundsen and Bergman (2007) have studied the functioning of Nordic electricity markets. During the second part of the 1990s, Denmark, Finland, Norway and Sweden created a unique multinational market for electricity, which meant that national electricity markets were opened up to cross-border trade and a common power exchange, Nord Pool, was established.[23] One of their conclusions is that there is now an integrated Nordic wholesale market and that this has effectively diluted the market power from national contexts. However, the retail markets have not integrated, at least to the same degree. Retail prices and trade margins differ significantly because of national legislation and structural differences. When studying Norway and Sweden,

Amundsen and Bergman found that in the case of Sweden three major generating companies – Vattenfal, Sydkraft and Fortum – have managed to become the dominant players in the retail market. Their market power is based on a policy of vertical integration in generation and retailing. One precondition for this is that the separation between retailing and distribution is required only in Sweden. The scholars explain that the competitive advantage of generation-retailing integrated companies probably stems from the lack of efficient markets for hedging against area price and quantity risk and that there are very significant entry barriers in generation. Even though there has also been a shift from a 'public service' to a 'business and profit' perspective among state-owned companies, their criticism is oriented towards problems related to legislation. The intention behind the legal separation of distribution and retailing was to stimulate competition by preventing cross-subsidization, but it opened up vertical integration between generation and retailing, which has in practice led to a reduction in the level of retail competition. Their conclusion is that from the point of view of households there seem to be two national rather than one integrated Norwegian–Swedish retail electricity market. They continue by stating that this also applies in the cases of both Finland and Denmark. Still they admit that to some extent Finland and Denmark are exposed to competition from foreign companies such as Vattenfall and E.ON. Thus they take the view that in future it will also be necessary to increase the market area for the retailers not only between the Nordic countries but also nationally and locally.

One example of creating markets for 'green' electricity can be seen in the Netherlands, where the government has made regular interventions in markets, demonstrating fundamental shifts in terms of both policy and approach. Van Rooijen and van Wees (2004) have studied the development of renewable energy policy making in the Netherlands. They characterize Dutch green electricity policy by dividing it into three phases: in the early 1990s, the government negotiated voluntary agreements with the energy distribution sector in relation to targets for green electricity sales, which were never met. In the second half of the 1990s, a regulatory energy tax was introduced, from which customers of green electricity were exempt. This led to a substantial increase in demand, which was met largely by green electricity imports, and did not lead to the development of additional domestic renewable energy capacity. Finally, in the third phase in 2003 there was a change in policy which shifted the focus from the promotion of demand to the promotion of supply through a system of regulated feed-in tariffs. Their conclusion is that, despite the renewable energy policies, the growth of the renewable energy market in the Netherlands has been small and targets have not been met in full. The share of renewable energy produced in the Netherlands in 2001 was 2.8 per cent of electricity consumption. The reasons for this are many, but, for instance, they state that the ambiguity of the policy has not created conditions for confidence among market partners. In contrast, countries such as

Denmark, which derives 21 per cent of its electricity from wind power, Spain, which obtains 7 per cent from wind power, and Germany, which achieves 6 per cent of its electricity from wind power, have succeeded in increasing the share of this alternative energy source, whereas France has invested in nuclear power so that by 2003 nuclear power accounted for a remarkable 80–85 per cent of France's total electricity production (Toke 2005, 48; van den Hoven and Froschauer 2004).

The power to choose between different energy options

One crucial question for the construction of a sustainable energy economy is how societies choose a suitable energy technology strategy from among the array of different options. In light of the convincing number of studies we suggest that even though energy policy is under constant pressure to change, the fundamental nature of national energy policy seems to be relatively stable. There are, of course, developments, collectively binding agreements, events and crises which force governments and other actors to reconsider their current policies, but in general this policy sector is fairly stable in comparison with other sectors. The determining forces are, for instance, those historically agglomerated social organizations, material infrastructures and mental schemes. Recent studies have revealed how solid the structures of the energy sector are and how historical choices have almost petrified the trajectory of energy sector development. To bring some fundamental changes into the energy sector is a laborious task, which requires strong governance, united policy advocacy coalitions and purposeful lobbying instead of wishful thinking and trusting to the benevolence of the main actors.

Hisschemöller, Bode and van de Kerkhof (2006) studied the relationship between technologies and institutions on the assumption that energy options may, in addition to the specific characteristics related to technology, costs and the public perception of risk, have some institutional requirements. Their arguments are that: (1) political institutions show a preference for some specific technologies and a dislike of others; (2) viable technologies do not conquer the market by virtue of their advantages for individual consumers and society at large; (3) socio-technological regimes control the choices of suitable technologies; and (4) breakthroughs in the field of renewables are believed to benefit from incentives for collaboration, especially the formation of new networks between vested interests and newcomers, but institutions still provide formal and informal rules that articulate both competition and collaboration in a specific way. Their conclusion is that the breakthrough of decentralized electricity production in combination with hydrogen needs major government interventions, not any kind of low profile governance or governance by corporate business. They are realists, as they state that this option is likely to be resisted by vested interests in the energy system, which in itself shows that it is in a position of disadvantage vis-à-vis other options and not only or primarily because of its technical or economic disabilities.

A more realistic picture of the formation of energy policy would acknowledge the substantial role that interest groups also play as active participants in the process. Lobbying the Parliament, politicians or authorities is a fairly conventional and accepted means of promoting the interests of certain actors. For instance, Markussen and Svendsen (2005) analysed how effective the lobbying of main industrial stakeholders was in a 2003 scheme for GHG emission allowance trading within the European Union. They focused only on the industrial groupings, as environmental and consumer organizations tend to be weak and to have limited lobbying power in the political arena. According to the literature to which they refer, the environmental interest groups are of only minor importance in relation to EU policy making (Michaelowa 1998; Daugbjerg and Svendsen 2001; Andresen and Gulbrandsen 2005; Gulbrandsen and Andresen 2005).

The ideology of the liberalization of energy policy and the freeing up of energy markets has suffered severe blows as nation-states have begun to protect their national interests to an increasing degree as the result of the instability of the energy markets. France is often seen as providing an example of very strong state intervention. In France, there are two dominant energy enterprises (the state-owned firms Electricité de France (EDF) and Gaz de France (GDF)), nuclear power is the main source of electricity and the ideology of energy policy is what Meritet (2007; see also Matláry 1997) calls 'public service'. She describes French energy policy in the context of the EU and refers to criticism in which it has been called the 'black sheep'. Even though there is considerable diversity in terms of energy policy issues among EU's Member States, there is this vision of the creation of a single energy market. At the same time there is also pressure to diversify the energy supply, lower the level of regulation and increase the share of renewable energy sources. The central notion of the establishment of a single competitive market has been a shock to the French culture of state intervention. According to Meritet, changes in the electric and gas industries, but also the reconstruction of network industries, are politically sensitive issues in France. European regulations imply that there should be a complete separation between competitive activities (generation, purchase and supply of gas and electricity) and regulated activities (transmission). The directives of 1996 (electricity) and 1988 (gas), which initiated the deregulation process, and the directive (2003), which aims to achieve the total opening up to competition, have brought about strong opposition from all parts of the French Parliament. Meritet (2007) concludes that France must consider how it can be part of the European process while simultaneously protecting its national interests. It seems obvious that nuclear power is an answer to the country's energy needs, the challenges of climate change and fears of energy supply disruption, even though sceptics counter that it is too costly and dangerous to be viable. The political elite have to face the challenge of the liberalization of energy markets within the context of domestic public intervention policy.

There are also a number of radical scholars who claim that for environmental and, in particular, climate reasons, the energy sector needs a period of 'creative destruction' (see, for example, Jacobsson, Andersson and Bångens 2002).[24] They are seeking the development of renewable energy technologies to replace the use of fossil fuels. They argue that a transformation of the energy sector requires the emergence and growth of new technological systems (see, inter alia, Jacobsson and Johnson, 2000) based on a range of renewable energy technologies, such as wind turbines, solar collectors, biomass-based combined heat and power plants and solar cells. There are many obstacles to the development of renewable energy technology, but Jacobsson et al. (2002) argue that governments have to offer support to the new technologies as they attempt to challenge technologies that have had several decades to mature. They base their views on the cases of natural gas and nuclear power, which have demonstrated that support may be needed for decades rather than just a few years. Their analysis of the evolution of the German technological system for solar cells prompts three conclusions: (1) An important role is played by organizations that articulate underlying values in favour of solar cells and legitimate the new technology. This work contains a repertoire of action, such as influencing the regulatory framework so that markets can be formed and coalitions of actors in favour of the new technology built. (2) The technology takes a long time to mature. They speak of the 'learning period', which in the case of solar cells has lasted several decades (defined as starting with the Research, Development and Demonstration (RDD) programme in the mid-1970s and ending when the cost of solar cells is somewhat closer to that of other power technologies). (3) The third conclusion is that, as the result of the nature of 'emerging' technology, there are great uncertainties over the potential performance and cost of the technologies.

In another study, Jacobsson and Lauberb (2006) have explored the reasons for the particularly rapid spread of two technologies in Germany: wind turbines and solar cells. They attributed this diffusion to the nature of the policy instruments employed as well as to the political process which led to the adoption of these instruments. The analysis demonstrates how the regulatory framework is formed in a 'battle over institutions', where the German Parliament, informed and supported by a growing advocacy coalition, backed support policies for electricity sourced from renewables in the face of often-reluctant governments and opposition from nuclear and coal interests. It also demonstrates that this major political and environmental achievement carries a modest price if we consider the total costs to society, i.e. including both subsidies to coal and the negative external economies of coal. The success factors in the German case were: (1) institutional changes; (2) market formation; (3) the formation of technology-specific advocacy coalitions; and (4) the entry of firms and other organizations. When dealing with the first factor, Jacobsson and Lauberb refer to the 'economics of innovation' literature. The institutional change requires a list of different measures, such as a redirection of science and technology policy well in advance of the emergence of markets, developing the value base (as it influences demand patterns), market regulations, tax policies as well as immediate and detailed practices. Second is the ability to generate markets for the new technology, which may involve the formation of standards (such as the Nordic telecommunication operators' decision to share a common Nordic Mobile Telephony (NMT) standard for mobile telecommunications), exploring niche markets (markets where the new technology is superior), a government subsidy and creating a 'protected space' or 'nursing market' for the new technology. Third, the scholars emphasize the importance of a broader constituency behind a specific technology. Their idea is that the development process should include not only individual firms, and related industry associations, but also a range of organizations, such as universities and also non-commercial organizations (like Greenpeace). Fourth, the entry of new firms is crucial to the transformation process, because new knowledge, capital and other resources are required – for example, to strengthen this technology-specific advocacy coalition and to legitimate the claims for the institutional set-up.

Nuclear power as a tempting option

In their Strategic Plan for the period 2005–09 the OECD Nuclear Energy Agency states that important changes have occurred in the energy and nuclear landscapes in recent years. They argue that at the same time as energy use continues to grow inexorably and fossil fuels continue to dominate the energy mix, serious concerns remain in respect of the security of supply, investment in the energy infrastructure and the threat of environmental damage caused by energy production. Another very serious challenge to the energy policy is climate change; at current rates of development, by 2030 CO_2 emissions are projected to be 70 per cent higher than today's levels. OECD member countries must therefore continue to favour energy savings as well as to promote the development of renewable and less CO_2-producing energy sources. In addition, many countries consider nuclear power to be a realistic option as part of their energy mix, because some energy policy actors consider that it provides significant environmental benefits, in particular in limiting CO_2 emissions and helping countries to fulfill commitments under the Kyoto Protocol.

The construction of a sustainable energy economy is a considerable challenge for countries throughout the world, in the face of the growth in worldwide energy demand. With no change of policy the energy needed in 2030 is expected to be 60 per cent higher (IEA, 2004). This increased demand will be met largely by fossil fuels, which will in turn become more expensive and increasingly scarce in future. Bleischwitz and Fuhrmann (2006) summarize the two most obvious challenges for energy systems: firstly, the world needs to ensure the security of energy supply at affordable and also foreseeable prices; and secondly, the world cannot ignore the environmental concerns and problems, which have been increasing steadily over recent decades.

It is unsurprising that under such circumstances some countries have shown a renewed interest in nuclear energy and have taken decisions to prepare for its future development. The construction of a new NPP has been decided on in Finland, although the competitiveness of new nuclear plants has not been demonstrated everywhere in today's market conditions and the reduction of their capital costs is a real challenge. By contrast, some countries have decided to co-operate in order to prepare a new generation of nuclear energy systems, including power plants and the associated fuel cycle facilities. The countries are forced into this kind of co-operation because they perceive the need for replacing older plants in the future (OECD 2005, 3).

Concerns about climate change and the security of supply have not been the only factors to open the door for the expansion of nuclear power. Another less publicly debated reason is the age of the current NPPs. A large number of NPPs are approaching the end of their initially estimated lifetimes and require either license extension or decommissioning. This can be seen from the report on the decommissioning of NPPs by IAEA (2004). The report concludes that 107 NPPs have been shut down or are undergoing decommissioning and that another 14 have been decommissioned already. The number of NPPs worldwide exceeds 500 units, with more than 400 units still in operation (IAEA 2004, 6). In 2002, nuclear power provided around 16 per cent of the world's electricity, with 441 units operating in 30 countries (IAEA 2003). The report states that it has been estimated that there are thousands of facilities worldwide that will eventually require some degree of decommissioning.

Another obstacle to constructing more NPPs has been the disposal of nuclear waste. As Elliot (2003, 449; see also Lowry 2007) states, the problem has not been just a question of the economic costs of nuclear waste disposal, it is also a matter of the longer-term viability of the nuclear option in terms of both safety and security. Some nuclear wastes remain dangerous for thousands of years. In some cases there has been progress in this previously unsolved problem. For instance, the Olkiluoto disposal facilities in Finland, and the identification of two sites in Sweden, are examples of how the nuclear industry and governments have developed approaches to the disposal of spent fuel in circumstances of rather low public and political confidence, even though in international comparison the trust in authorities, experts and also in industry is rather high in both of these countries. Finland and Sweden intended to solve this problem by encapsulating high-level nuclear wastes and storing it all underground in the bedrock. Both countries have carried out rather extensive public consultation at the local level on this issue (on the Finnish case, see, for example, Hokkanen 2001, 2007; Litmanen 2008). The construction of Finland's final disposal facility is scheduled to start in the 2010s and the facility should be operational after 2020. The deep repository model adopted in both countries has not convinced everyone, but at least in Finland there is one community which is willing to host it. In Chapter 6 Matti Kojo describes the political struggle that resulted in the positive decision at the local level in Eurajoki municipality.

We can say that nuclear power as an option for generating more electricity than previous NPPs produce has been in decline over the past two decades or so. To date this option has not been realistic for a large number of governments. The present-day situation is characterized by some dualism or ambiguity. Even though some countries have decided to phase out nuclear power because of concerns about severe accidents or problems with the management of radioactive waste, others have extended the licences of earlier installations or allowed the power upgrading of some plants. Governments and enterprises have become conscious of NPP closedown and the fact that the cost of electricity generated by existing NPPs is still competitive. For such reasons the number of NPP closures has been less dramatic than expected. For instance, despite regular increases in the demand for electricity, the contribution of nuclear power to electricity production in the OECD countries has remained stable – at approximately 24 per cent. From the global perspective the share of nuclear power global electricity generation has been steady for many years – at around 16 per cent. The IAEA secretariat, which produces a comprehensive Nuclear Technology Review every two years and shorter updates in the intervening years, reports that in 2004 five new NPPs were connected to the grid (two in Ukraine and one each in China, Japan and the Russian Federation), and one inactive plant was reconnected in Canada. This compares to two new grid connections (and, in Canada, two reconnections) in 2003 and six new grid connections in 2002 (IAEA 2005, 4). By contrast, five NPPs were closed down in 2004 – four 50 MW(e) units in the United Kingdom and the 1,185 MW(e) Ignalina-1 reactor in Lithuania. This compares to six retirements in 2003 and four in 2002 (IAEA 2005, 4). The report states that for the eighteenth consecutive year nuclear generation has continued to grow at the same pace as the overall level of global electricity consumption (IAEA 2005, 1). For instance, in 2002, nuclear power provided about 16 per cent of the world's electricity, with 441 units operating in 30 countries (IAEA 2003). The five new plants that were connected to the grid balanced the number that had retired. However, there were only two new construction starts in 2004, and in accordance with existing nuclear phase-out policies, the Obrigheim reactor in Germany, and Barsebäck-2 in Sweden, were shut down in May 2005.

Ambiguity is an appropriate term to describe the European Union's energy policy and its relationship with nuclear power. As both Matláry (1997, 14) and Frogatt (2007, 171–2) state, it is something of a paradox that energy policy in the EU has traditionally been rather insignificant, even though two of the three original treaties, the European Coal and Steel Community (ECSC, which came to existence in 1952), and the Euratom Treaty (signed in 1957), were concerned with the issue of energy. Today there is a shift towards internal energy markets, liberalization and deregulation, but this is far from being a completed process as different member countries have vested interests. Although the EU has 147 nuclear reactors in operation in 13 of the 25 Member States and they generate around one-third of the Union's electricity, some countries – such as France and Finland – are more enthusiastic about this source of energy than some others.

The recent turmoil in the international energy markets has increased the importance of energy policy in the EU. This new determination to shape the EU's energy policy means that Finland, as a member of the EU, is obliged to increase the share of energy from renewable sources in final consumption from the 2005 level of 28.5 per cent to a target share of 38 per cent in 2020. This national target is part of the climate change and energy package published by the European Commission in January 2008 (EU 2008). It includes a proposal for a framework directive relating to an increase in the share of renewable energy. According to a March 2007 European Council decision, the target is for the share to be increased to 20 per cent by 2020. This overall target differs from Member State to Member State.

According to Finnish industry, the strategy towards the development of a low carbon society should be combining nuclear power with renewables. One of their messages is that Finland can be regarded as one of the leading countries in the world in the use of bioenergy. Although the Confederation of Finnish Industries (EK) states that the use of bioenergy could be increased still further, they want to ensure the sufficiency of biomass as a raw material for the forest, food and chemical industries. Therefore their overall conclusion is that 'Taking into account the current, very high use of renewable energy sources in Finland, the national target of 38% is unrealistic' (EK 2008) They continue by stating that:

> Nuclear power is a form of energy that does not produce emissions. Alongside energy efficiency and the use of renewables, nuclear power plays an important role in controlling carbon dioxide emissions in Finland. Finland's carbon dioxide emissions were reduced considerably when nuclear energy was introduced in the late 1970s. Without nuclear power Finland's annual carbon dioxide emissions would be up to 20 million tonnes more. The introduction of a sixth nuclear reactor in 2011 will reduce annual emissions by 8 to 11 million tonnes. The low operating costs of nuclear power make it ideal for the production of basic energy. (EK 2008)

The Finnish commitment to nuclear energy is based in part on the fact that the country is part of the Nordic common electricity markets.

The Nordic countries have one of the highest proportions of electricity-intensive industries, including a paper and pulp industry, a steel industry, ferro alloys and aluminium smelting plants. These industries have been able to expand because of relatively low electricity prices. The power price-sensitive industry is closely interwoven with the energy industry and has a great deal of interest in the direction of government environmental policy. This conglomeration of industry is rather reluctant to invest in renewable energy sources and the governments are eager to continue the economic growth. Therefore the increasing use of nuclear power seems to offer a shortcut to a low carbon society without causing too much societal concern.

Determinants of recent Finnish nuclear decisions

This book is a collective effort by Finnish scholars of energy policy issues. The aim is to contribute to the global debate on nuclear power by shedding more light on the country's nuclear energy policy than has been undertaken previously. The analysis reveals the complexity of the country's energy policy and the determinants of recent nuclear power decisions.

Chapter 2, by Ari Lampinen, focuses on how the application for a new nuclear reactor was justified by the energy company TVO. His reading of the justification arguments is crucial in assisting our understanding of the favourable decision as all of the arguments were repeated in the government decision and in the process of parliamentary approval. The critical assessment of those four categories of arguments – namely security of energy supply, climate change and other environmental issues, cost of nuclear electricity and employment benefits – is based on the well-known fact that the Ministry of Trade and Industry has historically been a key player in Finnish nuclear power issues. Lampinen sheds light on how the MTI actually rules over energy policy as it exercises preparatory power, regulatory power, licensing power, energy policy expertise power, research and development power, environmental impact assessment power, energy market authority power and shareholder/ownership power in energy companies and in the transmission grid company. He points to the fact that during its preparation for government and parliamentary decision making, the MTI intentionally restricted energy political scenarios to nuclear and fossil fuel scenarios. In point of fact, the MTI has never included a renewable energy scenario in any of its alternative scenarios. This kind of analysis of justification arguments reveals clearly how one state actor can exert influence over both energy and nuclear policy.

Chapter 3, contributed by Harri Lammi, presents a view of the role of the anti-nuclear movement in the nuclear power debate and the decision-making process. He argues that in explaining the success of the pro-nuclear camp in 2002 we have to analyse the internal development of the anti-nuclear movement and the changing framing of nuclear power in public debate. As a representative of an anti-nuclear organization Lammi took part in the

2000–02 nuclear debate both in public and in Parliament. As a result he is able to describe the role and dynamics of anti-nuclear organizations and their lobbying strategy. His overall argument is that during the previous nuclear power debate of 1991–93 the public and the politicians were more receptive to the messages of the anti-nuclear movement. In these consequences, it was easy for a movement to find allies and to form coalitions. This previously heated nuclear debate crippled the movement as it was perceived that they had won the battle and that there was no need to continue anti-nuclear work. Lammi states that when the new nuclear debate began around 2000, the anti-nuclear movement had lost many members, member organizations, allies and, even more importantly, many capable older activists. Because of this discontinuity, they had to build a new anti-nuclear movement and to rethink the strategy. For a new generation of activists the choice was to leave behind the former antagonistic strategy, which had placed too much emphasis on the risks of nuclear power and did not fit their reformative ecological democracy ideology. The main focus of the campaign was the importance of renewable energy. They tried to promote the idea that renewable energy should play an important role in the fight against climate change and that the further construction of NPPs would have an injurious effect on the development of renewable energy technology. Another dimension in their campaign was the economic benefits that would arise from investing in renewable energy and what they saw as the economic madness of new NPPs. According to Lammi, in 2000 and 2001 there were also some preliminary decisions which need to be taken into account. As Parliament made the DiP on the SNF disposal plans and accepted the climate strategy prepared by the MTI which had only featured nuclear power and natural gas scenarios, the anti-nuclear movement lost powerful instruments. The nuclear waste problems were regarded as having been solved and renewable energy and energy efficiency were both deliberately excluded from the 'reasonable' national climate strategy. In addition, the Greens accepted the nuclear waste plans, even though they did not accept the further construction of nuclear power.

In Chapter 4 Annukka Berg takes a closer look at the decision making in the Parliament of Finland. Berg interviewed 12 members of Parliament approximately one year after the decision on the fifth NPP unit took place on 24 May 2002. The interviewees chosen were parliamentarians who had been potentially influenced by the nuclear discussion and the parliamentary decision-making process preceding the vote. (This included MPs who were undecided about nuclear power, MPs who had changed their voting behaviour compared to earlier nuclear votes and also MPs who had simply voted differently from their peer group.) In addition, her data also include interviews with two senior parliamentarians, both of whom have been influential figures in the opposing camps of the nuclear discussion. Thus the study considers a balanced sample of MPs, half of whom voted to construct the new NPP and half of whom voted against. Berg's theoretically driven analysis indicates that some of the MPs' discourses were consonant with the logic of either simple modernization or reflexive modernization. The former highlighted the minimal risk and economic benefits of nuclear power while the latter dealt with a critical stance towards the modern project. More intriguing than these rather straightforward 'for' and 'against' discourses, however, were the two other discourses. Discourses under the rubric of ecological modernization referred either to the idea that nuclear energy is an efficient means to combat climate change without jeopardizing economic development or to the idea that favouring nuclear power does not solve economic, technological and environmental problems, which would be much better addressed through introducing structural changes in energy policy and with renewable energy sources. Berg's analysis also reveals how the decision to declare the voting on the NPP an issue where 'freedom of conscience' would prevail strengthened MPs' positions and weakened the role of the political parties and parliamentary groups. During the parliamentary process MPs took part in the operation of different committees, organized hearings and discussions. Expert hearings in the committees and messages from the interest groups and other peer groups were deemed to be important, even though the contradictory conclusions of some experts forced them to assess the given information very carefully. Generally, Berg states that while the views of independent experts were much appreciated, the messages from influential interest groups such as the Central Union of Agricultural Producers and Forest Owners (MTK), the Central Organization of Finnish Trade Unions (SAK) and the Confederation of Finnish Industry and Employers (TT) – all of whom were in favour of the fifth NPP unit – received very careful consideration. One of her more surprising findings was that representatives of environmental organizations had difficulties in being granted the status of experts in parliamentary hearings. Some parliamentarians reported that they were cautious about accepting the messages of environmental organizations because their argumentation was perceived to be one-sided. Berg points out that the anti-nuclear movement as a whole faced great challenges in the discussion because of the structural changes in the anti- and pro-nuclear discourses and the altered pro-nuclear lobbying strategy that emphasized the importance of environmental issues and the role of nuclear power in dealing with the problems of climate change.

In Chapter 5 Erika Säynässalo's comparative analysis of Finland, France and Sweden indicates that when trying to understand national nuclear policy we have to pay attention to state institutions and their changing roles in the context of constitutions. For instance, in explaining Finnish nuclear power policy it is not sufficient to refer to how the strong state structures in Finland have contributed to the expansion of the nuclear power industry. Instead, her elaboration of strong versus weak state discussion leads to the conclusion that more sophisticated instruments have to be used than are offered by this distinction. Her suggestion is that in order to gain an insight into the three Western European states under consideration what is more appropriate is to consider the distinction between strong administrative states and strong party states. In a strong administrative state the constitution provides for the protection of the state's key industries from the hasty political reforms of political parties. By contrast, in a strong party state the constitution offers more opportunities for political forces to challenge previous economic and industrial policies when they have gained a majority in Parliament and a position of power within the government. Säynässalo's analysis indicates how in these three states the characteristics of the state are based on the constitution and provide different kinds of prerequisites for the complex interplay between political parties, the nuclear industry and the administration.

Säynnässalo argues that France and Finland can be viewed as strong administrative states. According to her, in both of these states the alliance between the state and the nuclear industry is guaranteed through a combination of constitutional and administrative procedures. Furthermore, in both of these states constitutions have been rather hostile to the idea of executive power in the hands of Parliament and, consequently, in neither of these states have the political parties been able to shape nuclear energy policy effectively. According to Säynässalo, in Finland, however, the centralization of power in the hands of the administration has followed a more complex logic than is the case in France. She points out that while the Finnish constitution has restricted the possibilities of Parliament to exert executive power in general, the paradox is that the Nuclear Energy Act (1987) offers members of the Finnish Parliament an exceptional opportunity to exert such power. Yet, in reality, this ability is strongly limited by the law. Parliament has only a kind of administrative power of veto when government hands down the Decision-in-Principle to Parliament. Therefore this administrative procedure laid down by the law limits the MPs' opportunities to affect the government's nuclear energy policy and instead offers to the administrative agencies prerequisites to govern the policy process. In general, she emphasizes that the weakness of Parliament parallels the inabilities of government parties to take a leading role in energy and nuclear power policy. She concludes that the binding role of legal-administrative procedure has contributed to ensuring that the nuclear industry has the ability to protect its legal rights to engage in entrepreneurship. The procedure itself does not allow government to decide on the schedule for decision making in nuclear energy, but it does force the government to take into account the interests of the applicant for the NPP when considering a proper schedule. The legal-administrative procedure is so binding in Finland that political actors can only make decisions according to this procedure – in other words, the political agenda on nuclear power is not defined on the grounds of political premises.

Säynässalo concludes that while the nuclear policies in France and Finland have been administration-led, in Sweden the government parties have had the capacity to govern the nuclear process. She argues that it is largely due to the important role of parties and party politics in governing nuclear policy that pro-nuclear politics has been challenged by means of the anti-nuclear movement. In Sweden, nuclear policy has been an essential element of governments' programmatic politics, while in Finland nuclear policy has been decided separately outside party-based parliamentary politics.

Chapter 6 by Matti Kojo provides the readers with a view of how Finnish nuclear waste policy was formed and how it was possible to achieve the unique decision to site spent nuclear fuel in Olkiluoto's nuclear oasis in the Eurajoki municipality. It consists of three complementary parts: the development of a national nuclear waste policy; the development of the site selection process since the early 1980s; and gaining local acceptance through interaction with the Eurajoki municipality, the nuclear waste company Posiva, the nuclear power company TVO and the government. It was originally believed that the best method of dealing with spent nuclear fuel was to export it. IVO with its Soviet–western style NPP[25] was able to send the waste to Soviet Union and later to Russia until political pressure forced the Finnish government to change the law. At the end of the 1970s TVO tried to secure reprocessing agreements with a French company or with a British company, but the costs were considered to be too high and the MTI did not consider the elimination of the tender from the point of view of TVO's operating licence. This initiated a change in nuclear waste management policy. More effort was put into research and development work by both the nuclear power companies and the MTI. The government made a Decision-in-Principle (DiP) in 1983, which established the timetable for spent fuel final disposal on the basis of TVO's programme. In 1995 these two nuclear power companies formed a joint nuclear waste company, Posiva, to take care of nuclear waste management. IVO's option to send waste to Russia was discontinued because of the amendment to the Nuclear Energy Act. The challenge for nuclear power companies was to find a suitable place for the repository. Kojo calls the siting strategy pragmatic, although originally the official aim was to make a decision on the basis of systematic geological studies. His detailed analysis of the search for a suitable location illustrates how many different policy instruments were in use and how the whole process exemplified multi-level governance. There were many different actors and the actions took place in several arenas. For instance, Kojo's reports on the interaction between TVO and Eurajoki municipality or Eurajoki municipality and the government are intriguing. Eurajoki councillors were placed under considerable pressure by TVO. The local government was lobbied for a good taxation decision by the representatives of Eurajoki. The representatives of Loviisa, the other Finnish nuclear oasis, challenged the dominant position of Eurajoki to host both a nuclear waste repository and also a new NPP. Officially interpreted, it was a fair game with the right cards, but on closer examination one finds that there were several games at different tables, each employing different cards and rules.

In Chapter 7 Tapio Litmanen concentrates on the stable sociocultural factors and dynamic societal changes that have proved crucial when trying to understand Finnish nuclear decisions. His argument is that societal risk evaluation varies over time. The temporary nature of risk evaluation emanates from fairly stable cultural factors, but it is equally important to focus on dynamic changes in society. Drawing on Mary Douglas's cultural risk theory Litmanen first considers the cultural characteristics of Finnish society, and what might be the basic values behind nuclear decisions. The Finnish political system is a particularly stable one, based on people's trust in welfare state structures, but also in a high level of trust in both the authorities and other citizens. Citizens' faith in the ideals of enlightenment, the state, technology and bureaucracy has deep historical roots. These factors help to understand why Finnish politicians are able to make difficult societal decisions without leaving a majority of citizens feeling that they have been cheated. Consensual corporatism, in which the government, labour unions and employers' organizations co-operate in order to predict societal development and foster economic growth, is part of the efficient functional governance of society. Although these factors were already in existence in 1993 Parliament rejected the plans for the construction of further NPPs in Finland. In order to understand the favourable NPP decision in 2002 and the nuclear waste decision in 2001 more detailed societal analysis is needed. More emphasis is given to contextual factors and dynamic changes. For example, changing a negative decision to a positive one requires a variety of actors to be involved in the process. The analysis of pro-nuclear lobbying in this chapter reveals that inside Parliament a powerful group of members of Parliament was directing the lobbying operations. Aided by a largely sympathetic media they were able to dominate the public debate, whereas ten years earlier the media had been more favourably disposed towards the messages of environmental and anti-nuclear movements. The anti-nuclear movement did not have such influential allies among the scientists as had existed during the 1970s and 1980s. In addition, the societal role of science had changed and different stakeholders had invoked the authority of science to promote their aims. As the pro-nuclear camp adopted the concession strategy, the front lines of the previous nuclear debate became blurred. Pro-nuclear actors put their faith in environmental values and anti-nuclear actors praised the good economic prospects of alternative energy sources. At the national level the country went through a dramatic transformation process because the economic depression made it imperative to adjust to a new environment. The rapid structural changes in the economy were followed by improved international competition. The new competition strategy was named Fair and Courageous Finland, meaning that the country also dares to take bold decisions. The nuclear power decisions were made in these national circumstances, but it must be remembered that the actors were also strongly influenced by the international developments triggered by the events of September 11, 2001.

Finally, Chapter 8 by Matti Kojo contains a summary of the most important research findings, describing how Finland has reached a political solution to the problem of nuclear waste disposal. Overall, we can conclude that it is important to understand the most important factors behind the recent nuclear power construction decision and also Finnish nuclear power policy in general.

Notes

1. At the moment three companies are competing for a licence to build the sixth nuclear power plant. Finnish utility Fortum Corporation (51 per cent owned by the Finnish government and the second largest power company in the Nordic region), Fennovoima (a consortium of companies, including, for example, Germany's E.On AG, stainless steel maker Outokumpu Oyj, regional utilities and Swedish mining and smelting group Boliden) and Finnish utility Teollisuuden Voima ltd, TVO, (building Europe's first European Pressurized Reactor at Olkiluoto on the west coast) have submitted their applications for the Decision-in-Principle (DiP) to the Finnish government. The Parliament of Finland will process the DIP during 2010. Some officials and industrial leaders have said that Finland needs more nuclear power to meet growing energy demand and to face the challenges of transforming the country to a low carbon society. At the moment, there is political debate about whether or not permission should be given to all three or only for two or perhaps a single NPP project. The option not to licence any of these rivals has by and large not been on the political agenda.
2. The €3 billion Olkiluoto 3 project started in 2005 and has been plagued by delays and is now expected to be online in 2012. It was supposed to be online in 2009.
3. Malcolm Grimston is acknowledged to be a long-standing consultant, advisor, and supporter of the UK nuclear industry.
4. During the years the name has been changed on a couple of occasions. At the beginning of the twentieth century it was called the Agrarian League (Maalaisliitto) and in the 1960s it was called the Centre Party (Keskustapuolue).
5. Suomen Sosialidemokraattinen Puolue was originally founded as the Finnish Labour Party, but in the early twentieth century it was changed to the present form.
6. Kansallinen Kokoomus is a centre-right political party in Finland.
7. Fingrid is the operator of the national transmission network, which carries all electricity at a voltage equal or higher than 110 kV. It owns 13,600 kilometres of lines, which means that it owns nearly all of the transmission lines in Finland and all of the cross-border lines. It started in September 1997, when Imatran Voima Oy (IVO, nowadays Fortum) and Pohjolan Voima (PVO) merged their transmission assets. Fingrid is owned by IVO and PVO (both 25 per cent), the state (12 per cent) and by institutional investors (38 per cent), with no other interests in electricity business. According to Pineau and Hämäläinen (2000, 187), this type of ownership is distinct from that in other countries.
8. They analyse the structure of the electricity sector in Finland before and after the restructuring of 1995. They divide the electricity market into six levels – generation, coordination of sales, system of operation, transmission, distribution and supply – each of which represents a specific subgroup of the industry that can be structured in its own way. Their sophisticated analysis is complemented with four other dimensions – market type (from monopoly to competition), ownership (from private to governmental), horizontal integration and the vertical integration – that also define structures (Pineau and Hämäläinen 2000, 182).

9. In 1995 the Electricity Market Authority was created as an independent expert body under the Ministry of Trade and Industry. Its main tasks are to supervise transmission pricing and issue licenses for transmission operations.
10. His thorough analysis of Finnish energy policy at the turning point was based on 29 focus interviews. The aim was to ascertain the effects of the changes in guidance ideology from planning and regulation to more competitive energy markets on Finnish energy policy and energy sector actors, such as political decision makers, authorities, enterprises and interest organizations. He interviewed 29 members of the energy policy elite: nine from the energy production sector, including public energy companies, private energy companies, regional electricity production, electricity companies owned by foreign capital and municipalities; 12 from the organization sector, including energy production, municipal, industry, entrepreneur and environmental organizations; four from central government (ministries); and four from the political parties (energy policy experts).
11. Ruostetsaari (1998, 212) perceives the subsystems consisting of: (1) energy enterprises, plants and other energy sector companies; (2) interest organizations; (3) authorities or some parts of some governmental authorities; (4) political parties or fractions of political parties; (5) research institutes or parts of them; and (6) media.
12. Today this former agrarian party is called The True Finns (Perussuomalaiset). It is a modern version of the Finnish Rural Party with a political ideology centred on nationalism and Euroscepticism.
13. This party was called Suomen Kristillinen Liitto, but today it is the Christian Democrats (Kristillisdemokraatit). Its roots are in the Christian faction of the conservative National Coalition Party.
14. Ruotsalainen kansanpuolue represent the Swedish-speaking minority in Finland and its ideology is liberalism.
15. The ideology of Vihreä liitto is a mixture of Green politics, traditional centre-left ideology and criticism of conventional political thinking with the rejection of the classification 'left' or 'right'.
16. Vasemmistoliitto was founded in Spring 1990 by the merger of the Democratic League for the Finnish People, the Finnish Communist Party and the Democratic League of Finnish Women. Ideologically, the party wants to be associated with the 'New Left' and Green socialism.
17. The International Energy Agency (IEA) is an autonomous body which was established in 1974 within the framework of the Organisation for Economic Co-operation and Development (OECD) to implement an international energy programme. It carries out a comprehensive programme of energy cooperation among 26 of the OECD's thirty member countries. The Organisation for Economic Co-operation and Development (OECD) is promoting a policy, which aims, for instance, to achieve the highest sustainable economic growth and employment and a rising standard of living in member countries, while maintaining financial stability, and thus to contribute to the development of the world economy.
18. Even though Finland's energy policies are generally perceived to be advanced, balanced and sound, some criticisms have also been voiced. The Agency states that three areas should be given particular attention. 'The first is supply security. Continued government policies are needed to address this challenge, and we urge a somewhat more diverse and long-term approach. The remaining two key areas are energy efficiency and R&D ... arenas where longer-term policies can benefit

the country's energy situation. Investments and policy enhancements in both of these areas help to improve energy security as well' (IEA 2008).

19. Of the 40 industrial countries, Finland was ranked number 32 (see Chapter 2 by Lampinen in this book).
20. Lampinen (2000) gathered those six strategic arguments from the public debate and official documents: (1) Finland is far more advanced than other countries in energy efficiency and reducing CO_2 gases; (2) Industry and exports are too energy-intensive; (3) Because of being a sparsely populated country, traffic produces higher emissions than in other countries; (4) The climate is cold in Finland; (5) The reference year 1990 is disadvantageous to Finland; (6) Emissions are predicted to rise in Finland more than in the EU on average.
21. Electricity market liberalization includes a number of changes: the introduction of wider market competition to ensure customer's ability to choose their energy supplier, the regulation of the transmission system to ensure suppliers access to potential customers, the deregulation of prices, and the expansion of international trade in electricity (Pineau, Hira and Froschauer 2004).
22. The only exceptions are those markets that are reasonably well integrated with neighbouring Member States or those where the main suppliers have, in fact, been privatized and purchased by foreign companies.
23. This reform started in Norway 1991, continued in Sweden 1996, and in Finland 1998 and was completed by Denmark in 2000.
24. Also the studies by van Rooijen & van Wees (2006) and Ek (2005) indicate how the creation of renewable energy markets is a long and laborious task.
25. To get a NPP Finland had to consider the political realities of the time. It became clear that it was politically impossible to get a purely western construction, because of Soviet Union's political pressure to buy their technology. Politicians understood that some kind of cooperation needed to be carried out with the Soviet Union. In the end the Loviisa NPP became a project, in which West and East cooperated in the field of nuclear technology for the first time. The reactor, turbine, generator and other main components are from the former Soviet Union, but safety systems, control systems and automation systems are of western origin. The steel containment and its related ice condensers were manufactured using Westinghouse licences. The degree of domestic origin was approximately 50 per cent (Michelsen and Särkikoski 2005).

References

Amundsen, E.S. and L. Bergman (2007) Integration of Multiple National Markets for Electricity. The Case of Norway and Sweden. *Energy Policy* 35(6), 3383–94.

Andresen, S. and L.H. Gulbrandsen (2005) The Role of Green NGOs in Promoting Climate Compliance. In O. Stokke, J. Hovi and G. Ulfstein (eds), *Implementing the Climate Regime: International Compliance*. London: Earthscan, pp. 169–86.

Arts, B. and J. van Tatenhove (2004) Policy and Power: A Conceptual Framework Between the 'Old' and 'New' Policy Idioms. *Policy Sciences* 37(3–4), 339–56.

Bent, R., L. Orr and R. Baker (2002) Introduction. In R. Bent, L. Orr and R. Baker, *Energy: Science, Policy, and the Pursuit of Sustainability*. Covelo, CA, USA: Island Press, pp. 1–9.

Berkhout, F., A. Smith and A. Stirling (2004) Socio-technological Regimes and Transition Contexts. In B. Elzen, F.W. Geels and K. Green (eds), *System Innovation and the Transition to Sustainability: Theory, Evidence and Policy*. Cheltenham: Edward Elgar, pp. 48–75.

Bleischwitz, R. and K. Fuhrmann (2006) Introduction to the Special Issue on 'Hydrogen' in 'Energy Policy'. *Energy Policy* 34(11), 1223–6.

Blowers, A. (2007) Preface. In D. Elliot (ed.), *Nuclear or Not? Does Nuclear Power Have a Place in a Sustainable Energy Future?* Energy, Climate and the Environment Series. Basingstoke: Palgrave Macmillan, pp. xi–xix.

Daugbjerg, C. and G.T. Svendsen (2001) *Green Taxation in Question.* Basingstoke: Palgrave Macmillan.

EK (2008) Low Carbon Energy. Confederation of Finnish Industries, EK. (http://www.ek.fi/www/en/energy_climate/low_carbon_energy.php – accessed 24 June 2007.)

Ek, K. (2005) Public and Private Attitudes Towards 'Green' Electricity: the Case of Swedish Wind Power. *Energy Policy* 33(13), 1677–89.

Elliot, D. (1997) *Energy, Society and Environment: Technology for a Sustainable Future.* London: Routledge.

Elliot, D. (2003) The Future of Nuclear Power. In G. Boyle, B. Everett and J. Ramage (eds), *Energy Systems and Sustainability: Power for a Sustainable Future.* Oxford: Oxford University Press in association with The Open University, pp. 435–73.

EU (2005) *Report from the Commission: Annual Report on the Implementation of the Gas and Electricity Internal Market.* {SEC(2004) 1720}. Commission of the European Communities. Brussels, 5.1.2005. COM(2004) 863 final.

EU (2008) *Proposal for a Directive of the European Parliament and of the Council on the Promotion of the Use of Energy from Renewable Sources.* Commission of the European Communities, Brussels, 23.1.2008. COM(2008) 19 final, 2008/0016 (COD). (http://ec.europa.eu/energy/res/legislation/doc/strategy/res_directive.pdf – Accessed 24 May 2008.)

Grimston, Malcolm (2005) *The Importance of Politics to Nuclear New Build: An Examination of the Relationship Between Political, Scientific and Public Mindsets and Its Influence on Decision-making in the Scientific and Technical Field.* Chatham House Report. London: The Royal Institute of International Affairs. (http://www.chathamhouse.org.uk/publications/papers/view/-/id/320/ – visited 3 September 2008.)

Gulbrandsen, L.H. and S. Andresen (2005) Influence in the Implementation of the Kyoto Protocol: Compliance, Flexibility Mechanisms, and Sinks. *Global Environmental Politics* 4(4), 54–75.

Hisschemöller, M., R. Bode and and M. van de Kerkhof (2006) What Governs the Transition to a Sustainable Hydrogen Economy? Articulating the Relationship Between Technologies and Political Institutions. *Energy Policy* 34(11), 1227–35.

Hokkanen, P. (2001) EIA and Decision Making in Search of Each Other: A Case Study. EIA of the Final Disposal Nuclear Waste in Finland. In T. Hilding-Rydevik (ed.), *EIA, Large Development Projects and Decision-making in the Nordic Countries.* Nordregio, R2001:6. Stockholm: Nordregio, pp. 95–152.

Hokkanen, P. (2007) *Kansalaisosallistuminen ympäristövaikutusten arviontimenettelyssä.* (Public participation in environmental impact assessment) Acta Universitatis Tamperensis 1285. Tampere: Tampere University Press.

Holttinen, H. (2005) Optimal Electricity Market for Wind Power. *Energy Policy* 33(16), 2052–63.

IAEA (2004) *Status of the Decommissioning of Nuclear Facilities around the World.* International Atomic Energy Agency. STI/PUB/1201. Vienna: International Atomic Energy Agency.

IAEA (2003) *Country Nuclear Power Profiles. 2003 Country Profiles.* Department of Nuclear Energy. Division of Nuclear Power. Nuclear Power Engineering Section. (http://wwwpub.iaea.org/MTCD/publications/PDF/cnpp2003/CNPP_Webpage/countryprofiles/Finland/Finland2003.htm – visited 21 June 2005.)

IAEA (2005) *Nuclear Technology Review — Update 2005*. International Atomic Energy Agency. General Conference GC(49)/INF/3. Forty-ninth regular session. Item 18 of the provisional agenda (GC(49)/1) (http://www.iaea.org/About/Policy/GC/GC49/Documents/gc49inf-3.pdf – visited 16 November 2005.)

IEA (2004a) *Energy Policies of IEA Countries. FINLAND. 2003 Review.* Paris: IEA. (http://www.iea.org/textbase/nppdf/free/2000/finland2003.pdf – visited 14 November 2005.)

IEA (2004b) *World Energy Outlook 2004*. Paris: IEA. (http://www.iea.org/textbase/nppdf/free/2004/weo2004.pdf – visited 5 February 2009.)

IEA (2008) *Energy Policies of IEA Countries – Finland – 2007 Review*. (http://www.iea.org/w/bookshop/add.aspx?id=340 – visited 2 April 2008.)

Jacobsson, S., B.A. Andersson and L. Bångens (2002) *Transforming the Energy System – the Evolution of the German Technological System for Solar Cells*. Electronic Working Paper Series. Paper No. 84. SPRU, Science and Technology Policy Research. (http://www.sussex.ac.uk/spru/1-6-1-2-1.html – visited 10 March 2006.)

Jacobsson, S. and A. Johnson (2000) The Diffusion of Renewable Energy Technology: an Analytical Framework and Key Issues for Research. *Energy Policy* 28(9), 625–40.

Jacobsson, S. and V. Lauberb (2006) The Politics and Policy of Energy System Transformation – Explaining the German Diffusion of Renewable Energy Technology. *Energy Policy* 34(3), 256–76.

Kaivo-oja, J. and J. Luukkanen (2004) The European Union Balancing Between CO_2 Reduction Commitments and Growth Policies: Decomposition Analyses. *Energy Policy* 32(13), 1511–30.

Kara, M. and S. Tuhkanen (2002) Finnish Energy System. Reliable and Efficient. In M. Kara, R. Hirvonen, L. Mattila, S. Viinikainen, S. Tuhkanen and I. Lind (eds), *Energy Visions 2030 for Finland*, 2nd edition. Helsinki: VTT Energy, Edita, pp. 14–35.

Kojo, M. (2004) Yhteenveto (Conclusion). In M. Kojo (ed.), *Ydinvoima, valta ja vastarinta* (Nuclear power, power and resistance). Helsinki: Like, pp. 231–58.

Kyllönen, S. (2004) Ydinvoiman ilmastonmuutos (The Climate Change of Nuclear Power). In M. Kojo (ed.), *Ydinvoima, valta ja vastarinta* (Nuclear Power, Power and Resistance). Helsinki: Like, pp. 51–85.

Lammi, H. (2004) Tarinat kovasta ytimestä (The Stories Concerning the Hard Nucleus). In M. Kojo (ed.), *Ydinvoima, valta ja vastarinta* (Nuclear Power, Power and Resistance). Helsinki: Like, pp. 11–50.

Lampinen, A. (2000) Suomalaisen ilmastotalouden ilmiöitä. Miten energiansäästö ja muut win-win-mahdollisuudet tullaan ottamaan huomioon Suomen ilmastonmuutoksen torjuntastrategiassa? (Phenomena in Finnish Climate Economics. How Will Energy Efficiency and Other Win–Win Opportunities be Taken into Account in the Finnish National Climate Change Strategy?) *Futura* 19(3), 81–95.

Lipponen, P. (2007) The Future of Nuclear Energy in Europe. Speech given as by the speaker of the Parliament of Finland, Warsaw 15 January 2007.
(http://www.paavolipponen.org/cgi-bin/puheet.php?key=384 – visited 3.9.2007.)

Litmanen, T. (2008) The Changing Role and Contribution of Social Science to Nuclear Waste Management in Finland. *Energy & Environment* 19(3–4), 427–53.

Lovins, A.B. (2004) *Winning the Oil Endgame*. London: Earthscan.

Lowry, D. (2007) Nuclear Waste: the Protracted Debate in the UK. In D. Elliot (ed.), *Nuclear or Not? Does Nuclear Power Have a Place in a Sustainable Energy Future?* Energy, Climate and the Environment Series. Basingstoke: Palgrave Macmillan, pp. 115–31.

Markussen, P. and G.T. Svendsen (2005) Industry Lobbying and the Political Economy of GHG Trade in the European Union. *Energy Policy* 33(2), 245–55.

Matláry, J.H. (1997) *Energy Policy in the European Union*. Basingstoke: Macmillan Press Ltd.

Meritet, S. (2007) French Perspectives in Emerging European Union Energy Policy. *Energy Policy* 35(10), 4767–71.

Michaelowa, A. (1998) Climate Policy and Interest Groups — A Public Choice Analysis. *Intereconomics: Review of European Economic Policy* 33(6), 251–9.

Michelsen, K.-E and T. Särkikoski (2005) *Suomalainen ydinvoimalaitos* (Finnish nuclear power plant). Helsinki: Edita.

Morland, M. (2001) Climate Change and Nuclear Energy. *International Relations* 15(4), 53–64.

Ministry of Sustainable Development. 2005. Closing of Barsebäck 2 – background. (http://www.sweden.gov.se/sb/d/5745/a/45905;jsessionid=askdlzkMfxR8 – 10 November 2005.)

MTI (2005) Ministry of Trade and Industry. Energy. (http://www.ktm.fi/index.phtml?c=www&l=en&s=7 – 16 November 2005.)

Myllyntaus, T. (1991) *Electrifying Finland: the Transfer of a New Technology into a Late Industrialising Economy*. Basingstoke: Macmillan; Helsinki: Etla.

Nurmi, M. (1980) *Energiatalous* (Energy Economy). Helsinki: Weilin+Göös.

OECD (2005) *The Strategic Plan of the Nuclear Energy Agency 2005–2009. Summary*. Organisation for Economic Co-operation and Development 2005, Nuclear Energy Agency No. 6074. (http://www.nea.fr/html/nea/strat05-summary.pdf – visited 21 June 2005.)

Pineau, P., A. Hira and K. Froschauer (2004) Measuring International Electricity Integration: a Comparative Study of the Power Systems Under the Nordic Council, MERCOSUR, and NAFTA. *Energy Policy* 32(13), 1457–75.

Pineau, P.-O. and R.P. Hämäläinen (2000) A Perspective on the Restructuring of the Finnish Electricity Market. *Energy Policy* 28(3), 181–92.

Ruostetsaari, I. (1998) *Energiapolitiikka käännekohdassa. Järjestöt ja yritykset vapautuvilla energiamarkkinoilla*. (Energy policy at the turning point. Organizations and enterprises in the opening markets). University of Tampere, Department of Political Science and International Relations, Research Reports 8/1998.

SEPA & SEA (2004) Swedish Climate Strategy: a Basis for the Evaluation of Swedish Climate Work. Checkpoint 2004 – The Swedish Environmental Protection Agency and the Swedish Energy Agency. (http://www.stem.se/web/biblshop_eng.nsf/FilAtkomst/ET33_04.pdf/$FILE/ET33_04.pdf?OpenElement – 10.11.2005.)

Toke, D. (2005) Will the Government Catch the Wind? *The Political Quarterly* 76(1), 48–56.

van den Hoven, A. and K. Forschauer (2004) Limiting Regional Electricity Sector Integration and Market Reform. The Cases of France in the EU and Canada in the NAFTA Region. *Comparative Political Studies* 37(9), 1079–103.

van Rooijen, S.N.M. and M.T. van Wees (2004) Green Electricity Policies in the Netherlands: an Analysis of Policy Decisions. *Energy Policy* 34(1), 60–71.

van Tatenhove, J., B. Arts and P. Leroy (eds) (2000) *Political Modernisation and the Environment: The Renewal of Environmental Policy Arrangements*. Dordrecht: Kluwer Academic Publishers.

Winskel, M (2002) Autonomy's End: Nuclear Power and the Privatization of the British Electricity Supply Industry. *Social Studies of Science* 32(3), 439–67.

Part II
The Application and Resistance

2

An Analysis of the Justification Arguments in the Application for the New Nuclear Reactor in Finland

Ari Lampinen

New nuclear power in Finland

In Finland nuclear energy is the only form of energy production that is governed by a specific law and liabilities: the law on atomic energy was enacted in 1957 and the law on atomic liabilities was enacted in 1972. Following the Chernobyl incident in 1986 a new law on nuclear energy, the Nuclear Energy Act (990/1987) accompanied by the Nuclear Energy Decree (161/1988), was deemed necessary and this has been in force since 1988. Under the act, new reactors larger than 50 MW_{th} require a government Decision-in-Principle (DiP) that the new facility will be of benefit to society. The DiP will also require parliamentary approval.

The four existing Finnish reactors were built in the 1970s, before the adoption of the Nuclear Energy Act, and a new, fifth reactor, Olkiluoto-3, is currently under construction. In the DiP application for the fifth reactor several justifications were advanced by the applicant company, TVO. Every one of these was accepted and repeated in both the government's decision and the parliamentary approval. The arguments fell into several categories, addressing issues such as energy supply, climate change and other environmental issues, the cost of nuclear electricity and the benefits to employment that will arise from the construction of a new reactor. During the decision process an extensive external expert review was conducted using invited and open statement calls by the government and expert hearing sessions in front of several parliamentary committees.

Although the Nuclear Energy Act enables in principle a very democratic decision-making process, the Ministry of Trade and Industry (MTI) – from January 2008 the Ministry of Employment and the Economy – has highly concentrated powers to ensure a smoother process than would be apparent to the external observer. The multiple roles of the MTI included acting as a regulatory and permission authority on one hand, and substantial owner of the applicant companies on the other hand. For the general public this conflict of interest has not been visible.

The experiences of this application cycle raise several policy questions relating to governmental and parliamentary roles and responsibilities. This appraisal is necessary when processing new DiP applications.

Proposals for new reactors have been made on various occasions since the four existing reactors began their operations in the period from 1977 to 1982. In the 1970s the state-owned power company IVO had plans for an additional 10,000 MW_e of nuclear power capacity. At the time of writing IVO (currently Fortum) has not realized any of that planned capacity, but TVO (with 25 per cent Fortum ownership) has proceeded successfully with Olkiluoto-3. Construction of the fifth nuclear reactor, Olkiluoto-3, began in 2005 following an application process that had lasted seven years. According to the DiP application, the new 1600 MW_e Franco-German EPR reactor (PWR) was supposed to start operating in early 2009 (but it will be delayed by at least three years) and produce 13 TWh_e annually – a figure that corresponds to 15 per cent of present national electricity consumption. Currently, there are plans in various stages for five additional reactors and two additional NPPs. In early 2007 the German power company E.ON began a process for a new nuclear power plant (NPP) near Loviisa NPP, but it failed to reserve land for the purpose following local resistance that arose mainly from the need to evacuate people out of the five-kilometre safety zone around the new NPP. It is not clear whether in future E.ON will proceed with its own NPP project or whether it will concentrate only on Fennovoima projects. TVO started an EIA process for Olkiluoto-4 in May 2007 and Fortum started an EIA process for Loviisa-3 in June 2007. A new Finnish–German nuclear power company, Fennovoima, was established in June 2007 with the intention of building new NPPs. The German power company E.ON is the largest shareholder in this new venture, holding 34 per cent of the shares. They started EIA processes in January 2008 at four sites with a view to building one NPP with either one or two reactors. Ruotsinpyhtää is located on the south coast of Finland near the existing Loviisa NPP, whereas Kristiinankaupunki, Pyhäjoki and Simo are located on the west coast north of Olkiluoto, some 200 kilometres apart. In April 2008, TVO submitted its new DiP application for Olkiluoto-4. The Fennovoima DiP application for one or two reactors was submitted in January 2009 for three instead of four alternative sites: Kristiinankaupunki is not included due to municipal resistance to the project. Fortum finished the EIA report for Loviisa-3 reactor in April 2008 and submitted its DiP application in February 2009. Government is expected to make decisions of all three DiP applications simultaneously during the first half of 2010.

The role of the Ministry of Trade and Industry (MTI) in Finnish nuclear energy policy

According to the provisions of the EIA Act and the Nuclear Energy Act, MTI is the responsible authority for environmental impact assessments, the DiP, and the building permit and utilization permit processes for new reactors. MTI, as the controller of state ownership, had a major role in all of the key companies involved in the fifth reactor process (in 2007 the state shareholding ownership was moved from MTI to the State Council):

- Fortum Corporation was created in 1998 through the merger of the state-owned power company IVO and the state-owned oil company Neste. The Finnish state sold 49 per cent of the shares via the Helsinki Stock Exchange where Fortum became listed in 1998. In 2005 the oil business was separated into a new Neste Oil Corporation and Fortum Corporation became a power and heat company. The state owns 51.1 per cent of Fortum and 50.1 per cent of Neste shares and voting rights. Fortum owns two of the four nuclear reactors in Finland: Loviisa-1 and Loviisa-2.
- TVO is a private–public partnership company established in 1969 for nuclear power production, and extended later to the areas of coal and wind power. Pohjolan Voima (PVO) owns the majority of the shares, Fortum holds 26 per cent and other state or municipal owners have more than 10 per cent. TVO owns two of the four nuclear reactors in Finland: Olkiluoto-1 and Olkiluoto-2.
- PVO is a private–public power company established in 1943 primarily for providing power for its shareholders, originally by hydropower. The shareholders include private and public companies and municipalities, with the majority of the shares being held by the pulp and paper industry. The state has a significant minority ownership through its involvement with paper and power industry companies.
- POSIVA is a nuclear waste company established in 1995 as a joint venture between Fortum (40 per cent) and TVO (60 per cent).
- The Fingrid transmission grid company was established in 1996 to manage the national transmission grid. The state owns 12 per cent of the shares (and 50 per cent of voting rights) and Fortum 25 per cent of the shares.
- Kemijoki Ltd is a hydropower company established in 1954, with 50.1 per cent state ownership. It provides regulating power for NPPs.

Most of the key authorities involved in nuclear power are administered under the MTI: these include the Energy Market Authority (EMV), the Technical Research Centre of Finland (VTT) and Finnish National Technology Agency (TEKES).

The radiation safety authority STUK and the municipalities hosting the reactors are the most important actors with no direct MTI control. STUK is a state organization administered by the Ministry of Social Affairs and Health which has the authority to establish safety norms. Its permission is required for the utilization of nuclear reactors.

MTI is in overall charge of Finnish energy policy and acts, either by itself or through VTT, as the main expert in the areas of energy policy, climate policy and energy technology-related affairs in the government and Parliament. MTI also controls, either directly or through VTT and TEKES, the direction of energy research policy and funding in Finland. This administrative-led parliamentarism is typical of Finland, as described by Säynässalo in Chapter 5. One example of a mechanism used for this purpose by MTI is the energy political scenario work undertaken to assist the government and Parliament in its decision-making processes. The alternatives have always been restricted to nuclear and fossil scenarios, as happened, for example, in the national climate strategy highly relevant to the fifth reactor decision process (MTI 2001). It is significant that in all the alternatives that it has proposed MTI has never included a renewable energy scenario. When the renewable energy scenario is included in long-term scenario comparisons, such as that made by the United Nations Intergovernmental Panel on Climate Change (IPCC 2000), it usually suggests a higher rate of GNP growth than occurs in those scenarios emphasizing non-renewable energy technologies. The same conclusion is also found in renewable energy scenarios made by Finnish universities and environmental organizations explicitly for Finnish conditions, but these have also not been included in the political and administrative decision-making processes. MTI has been able to create a hegemonic political discourse in which only one or two alternatives of its own choosing are actually considered to be relevant, although from the outside it may appear that a much wider set of alternatives is on the table. This phenomenon has been a characteristic feature of Finnish climate policy (Tirkkonen 2000).

MTI's energy policy is in conflict with EU-level energy policy. MTI has held the view that energy policy should be market driven, whereas the EU-level policy emphasizes active government roles in the promotion of renewable energy and energy conservation. This policy was further strengthened in 2007 by the Treaty of Lisbon and in January 2008 by the energy and climate policy directive package.

In the case of the fifth reactor, the initiative came from the paper and metal industries, and MTI used its extensive powers to make the process as smooth as possible. But EU directives requiring the development of active government policies to promote renewables and energy conservation have encountered significant resistance from MTI. One example is the RES-E Directive (2001/77/EC) on the promotion of electricity generation from renewable energy sources. Its draft was discussed in the Council of the European Union and the Finnish Parliament simultaneously with the fifth reactor process. The EU Commission, the European Parliament and the majority of governments in the Council of the EU required each Member State to set binding targets for the share of electricity production to be achieved from renewable energy sources. The target set for Finland was 35 per cent, which represents a five percentage point increase from the Kyoto Protocol base year 1990. The MTI convinced the Finnish Parliament that such a level was impossible to achieve and that no binding targets should be accepted. Since common consent was required in the Council of the EU, the target for

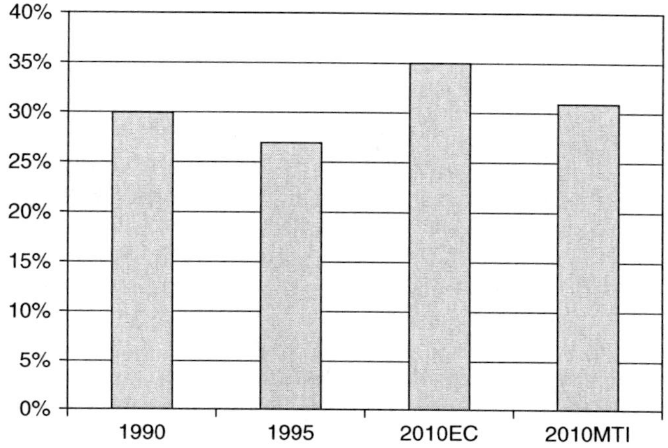

Figure 2.1 Share of renewables in power production in Finland in 1990 and 1995, the suggested EU binding commitment for Finland for 2010 in the RES-E Directive draft and the MTI voluntary target for 2010 in the final RES-E Directive (2001/77/EC)

Finland in the directive was downgraded to 31.5 per cent and the targets for all Member States became voluntary (Figure 2.1).

If this directive had come into force with the proposed mandatory targets, it would have required an increase of renewable electricity production by approximately the same amount as is produced by one existing Finnish nuclear reactor. It would have meant a serious conflict with the fifth reactor process. In 2004 the share of renewables in domestic power production was 29 per cent, originating from hydropower and the waste wood from the forest industry (Statistics Finland 2005), both of which are very competitive in power production without any state promotion policies (see the figures for wood cogeneration in Figure 2.5). The large share of bioelectricity (11.5 per cent) is regarded as an international success story, with only a minimal level of state involvement (Kautto 2005). However, in MTI this high share is seen as a valid reason for resisting an active government role in respect of renewables.

The strong lead provided by the MTI is also seen in their role as the responsible authority in nuclear EIA processes. Hokkanen and Ruuskanen (2005) studied the EIA process of permanent nuclear waste disposal and compared it with a road construction EIA case, where the responsible authority was the Central Finland Environment Centre, administered under the Ministry of the Environment. Regional environmental centres are the responsible authorities in all EIA processes except nuclear facilities. Hokkanen and Ruuskanen (2005) found that most of the comments given in the EIA review were value-based, but that a substantial number of fact-based statements were also given. There was a wide difference between these two cases in terms of how the inputs were taken into account. They were taken fully into account in the road construction case. In the nuclear waste disposal case 120 written statements were made in the open call. Of these, 45 included factual reviews and 21 included technical details. These were put forward by technical experts or used information produced by technical experts on the issue. The official statement of the MTI, which is binding on the applicant, omitted all arguments given in the open call except those relating to the siting of the facility and some of those that were in accord with the applicant's views or mandatory statements. The official MTI statement included those expert views that were submitted by organizations with mandatory statement rights conferred by the Nuclear Energy Act: STUK, five ministries, the Geological Survey of Finland (GTK), the Technical Research Centre of Finland (VTT), the Finnish Road Administration (Tiehallinto), the Finnish Environment Institute (SYKE) and affected municipalities. A few sentences giving an overview of each statement given in the open call was listed in an appendix that is not binding on the applicant. The general facts of the EIA application were not questioned in the MTI statements. Technical details were discussed on the basis of the mandatory statements and some further requirements were made for the applicant.

Overview of the application process for the fifth reactor

According to section 11 of the Nuclear Energy Act (990/1987) new reactors larger than 50 MW_{th} require a government DiP that the construction project is of benefit to society:

> Nuclear Energy Act: Section 11 – Decision-in-Principle
> Construction of a nuclear facility of considerable general significance shall require a Government Decision-in-Principle on that the construction project is in line with the overall good of society. Of the nuclear facilities referred to in section 3(5), those of considerable general significance are:
>
> (1) facilities operated for the generation of nuclear energy having a thermal power higher than 50 megawatts;
> (2) facilities serving as repositories for nuclear waste; and
> (3) facilities operated for purposes other than the generation of nuclear energy having, at a given moment, an amount of nuclear material or waste or involving a radiation risk, as defined by decree, that shall be deemed comparable with nuclear facilities as defined in paragraph 1.

In addition, parliamentary approval of the government DiP is required under section 15 of the Act:

> Nuclear Energy Act: Section 15 – Notification to Parliament and Parliament's decision
> The Government DiP, made under section 11, in which the construction of the nuclear facility is judged to be in line with the overall good of society,

> shall be forwarded, without delay, to Parliament for perusal. Parliament may reverse the DiP as such or may decide that it remains in force as such. Before Parliament has made its decision, the applicant shall not initiate measures to be laid down by decree which, because of their economic significance, might impede Parliament's, or the Government's, possibilities to decide at its discretion.

MTI is the responsible authority in the DiP application process.[1] Based on the provisions of section 16 of the Nuclear Energy Act nuclear reactors also require a building permit and utilization permit from the government (in practise the MTI). In November 2000 TVO made a DiP application based on the Nuclear Energy Act (TVO 2000) concerning two alternative sites for the reactor: Loviisa and Olkiluoto. Fortum did not submit a separate DiP application. The MTI organized a review process of the TVO application based on the Nuclear Energy Act and the Nuclear Energy Decree. A total of 148 statements were received from 152 organizations, groups and individuals. After analysing the application and the statements the MTI submitted the application to the government. The application was accepted by the government in January 2002 (VN 2002) and was submitted to Parliament for approval.

Parliament organized a very broad expert hearing process in which about 200 experts participated in eight parliamentary committees. It was supported by one of the largest social discussion and lobbying processes in Finnish history, including an NGO campaign that is described in Chapter 3 by Lammi. The external review organized by Parliament is a standard feature in parliamentary decision-making processes and not specifically required under the nuclear energy legislation. Because of the importance and difficulty of the nuclear power issue the review process became broader than might be the case in relation to other issues. As a result, a large amount of information was delivered, which offered an excellent basis for the formation of opinion. But, on the other hand, since all members of Parliament participated in selecting the invited experts, the process also helped to legitimize any existing opinions (as discussed in Chapter 4 by Berg).

On 24 May 2002 the government position was accepted by 107 votes to 92. The importance of the issue was shown in the results of the voting process: for the first time in the history of the Finnish Parliament, there were no absentees or abstentions.

Review of the justifications of the application for DiP

Unlike all other types of large thermal power plants, an environmental permit based on the Environmental Protection Act (EPA, 86/2000), the Finnish implementation of the IPPC (Integrated Pollution Prevention and Control) Directive (96/61/EC), is not the primary quality requirement in the case of nuclear reactors, since radiation-related issues of NPPs are not regulated by the EPA. The NPP environmental permit is merely meant for controlling impacts of cooling water discharges. The EPA has several provisions that enable better environmental governance than the Nuclear Energy Act, such as the requirement to use the best available technologies and the best environmental practices. Section 59 of the Environmental Protection Act allows for the revocation of an environmental permit if some of the information provided by applicant in the application turns out to be either inaccurate or false:

> Environmental Protection Act: Section 59 – Revoking a permit
> On the initiative of the supervisory authority the authority granting the permit may revoke it if:
>
> 1) the applicant has provided erroneous information that is material to the preconditions for granting the permit;...

By contrast, the Nuclear Energy Act does not allow the revoking of a DiP for any reason and it does not allow the refusal or withdrawal of the building permit or the utilization permit in case of inaccurate or false information in the DiP or the building permit application.

Furthermore, the environmental impact assessment, which must be made in accordance with the provisions of the EIA Act, does not need to be taken into account in any relevant decisions made under the Nuclear Energy Act. The Nuclear Energy Act requires only that the EIA has been made publicly available and the Nuclear Energy Decree requires that the EIA is included in the application for a DiP. There are no requirements to take the EIA into account in DiP, building permit or utilization permit decisions. On the contrary, in the building permit decisions regarding other types of large thermal power plants, the permission authority is required, under section 52 of the EPA, to take the EIA into account and to declare how it has been done. The same applies to hydropower plants under section 23a of the Water Act. In practice, information provided by the EIA is usually used in the permit ordinances.

For these legislative reasons NPP applicants have less incentive to provide accurate information than the applicants for other types of power plants. The main incentive is to provide a politically convenient application in order to sell the idea. Essentially, the Nuclear Energy Act shifts the responsibility for application data inaccuracies from the applicant to the government and Parliament.

In the TVO application for the DiP several arguments were presented to substantiate why the reactor would be beneficial for the society. The main arguments were:

1. Environmental benefits: Climate and environmental policies.
2. Energy security benefits: Security of electricity production and energy import independency.

3. Economic benefits: Competitive and stable price of nuclear electricity and employment effects.

Application review processes

The application was reviewed in 148 statements within a process organized by the MTI in 2001 and about 200 expert statements within a process organised by Parliament in 2002.

In 2001 the MTI organized a review process in accordance with the Nuclear Energy Act and the Nuclear Energy Decree. Compulsory statements based on section 12 of the Act were requested from STUK, the Ministry of the Environment, the municipalities of Loviisa and Eurajoki and their neighbouring municipalities. Compulsory statements were requested, as required by section 25 of the Nuclear Energy Decree, from the Ministry of the Interior, the Ministry of Defence, the relevant provincial and regional governments and environmental centres, and two government advisory committees on nuclear energy. A total of 29 statements were received in response to compulsory requests (VN 2002). All of these statements came from the public sector: 14 municipalities, nine provincial organizations, four governmental organizations and two governmental committees. The application was declared to be acceptable in 14 statements, including those submitted by the Eurajoki and Loviisa municipalities. At least one of the main arguments was criticized in 15 statements. The Ministry of the Environment offered criticism of all three main arguments.

STUK did not consider the safety features adequate in any of the reactor alternatives presented in the application. According to the Nuclear Energy Act, STUK is the authority that defines and monitors the safety of nuclear installations. Since the operation permit for reactors is dependent on STUK approval, the requirements of STUK must be met.

The MTI also requested statements from the Ministry of Agriculture and Forestry, the Ministry of Social Affairs and Health, the Ministry of Transport and Communications, the Ministry of Finance, the Finnish Environment Institute, the Federation of Finnish Energy Companies, the transmission grid operator Fingrid and the Finnish Association for Nature Conservation. A statement was also requested from the Government of Sweden. In addition, the MTI sent the application to eight governmental or non-governmental organizations and invited them to comment on the proposals. A total of 16 statements were received (VN 2002). These came from five governmental organizations, one company (Fingrid), eight non-governmental organizations and one Swedish governmental organization (nuclear safety authority SKI). The application was found acceptable as such in eight statements. At least one of the main arguments was criticized in eight statements. The Finnish Association for Nature Conservation criticized all three main arguments.

In addition, 91 organizations, groups and individuals sent their written statements to MTI based on an open call and 16 organizations and individuals gave their opinions orally in two public hearings organized by the MTI in Loviisa and Olkiluoto (VN 2002). The statements came from 48 non-governmental organizations, 36 individual citizens, 17 citizen groups, five companies and one political party. Some of the statements were given jointly by more than one organization. The application was found acceptable in 37 statements. It was not found acceptable in 61 statements for a multitude of reasons, both value and fact based. In the rest, no final position was given.

TVO gave the MTI feedback on the statements given (VN 2002). TVO agreed with STUK about the need to improve reactor safety: the required features were included in their tendering process and also the building permission application. TVO also reacted to several statements that considered the environmental impact assessment in relation to cooling water to be inadequate, especially in relation to nearby nature conservation areas. An updated version was sent to the MTI. TVO did not agree with other points of criticism.

In the MTI analysis used as a substantiation for the positive decision of the government, the STUK's criticism in respect of reactor safety was included to comply with the requirements of the Nuclear Energy Act. However, none of the other points of criticism was agreed to and no actions were suggested to deal with any of the issues raised. The justification arguments of TVO were repeated without doubts in the positive government DiP (VN 2002).

The parliamentary process included statements from some 200 experts to eight parliamentary committees. They represented ministries and other governmental organizations, local administrations, universities and other research organizations, companies, business organizations and environmental organizations. In addition, very broad public discussion also engaging members of the Parliament took place in the media. Hokkanen (2004) analysed the ten largest national and provincial newspapers over a 4.5-month period and found 136 written opinions, of which 56 per cent represented negative positions and 19 per cent positive positions towards the proposed new reactor.

Many of the points of criticism given in Parliament statements, in statements given for the MTI, and raised in public discussion were agreed to in Parliament. The approval of the government DiP was connected with demands for MTI and other government action. They relate to increasing renewable energy use, increasing energy efficiency and decreasing coal use. Details of the main arguments and their criticism are presented below.

Environmental benefits: climate and environmental policies

The most common criticism presented in both the MTI and parliamentary review processes was the possible negative impact of additional nuclear power on the promotion of renewable energy and energy conservation. This is closely related to the Finnish climate strategy in which the MTI had designed two alternative scenarios to tackle climate change: natural gas and nuclear

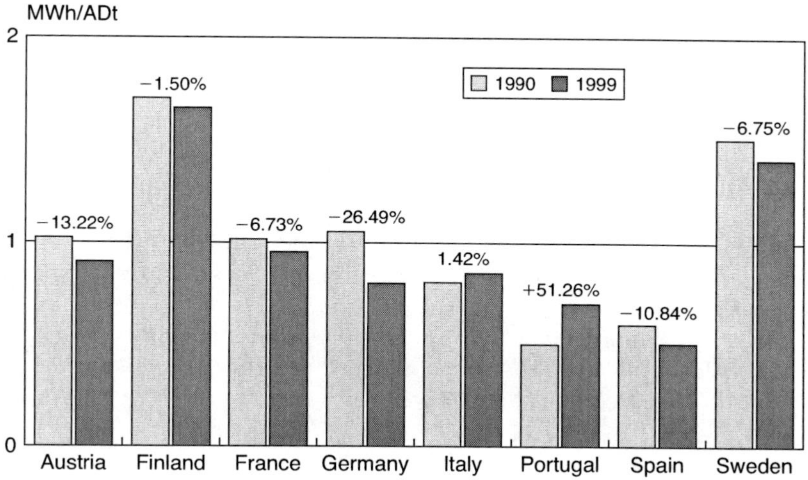

Figure 2.2 Specific electricity consumption in MWh for producing one tonne of air dry paper in European countries in 1990 and in 1999 (CEPI 2002). Finland has the most electricity intensive pulp and paper industry

energy scenarios, but no renewable energy scenario (MTI 2001). The positive DiP was interpreted by many proponents of nuclear power as a decision in favour of nuclear energy in future Finnish energy policy and climate strategy, leading not only to the construction of the fifth reactor, but also the prospect of further reactors. Therefore, nuclear power might fill the electricity market or create expectations of filling the market, leaving no incentives for renewable power investments. Experience of the impact of building the four existing reactors supports this fear. Not only did they freeze renewable power investments; they also created a need to increase power consumption, in both heavy industry and households. One example is the pulp and paper industry, which began a trend of moving from wood-intensive and net electricity-producing chemical pulping to electricity-intensive mechanical pulping which resulted in the least power-efficient paper manufacturing in Europe (Figure 2.2). In 2005 the pulp and paper industry consumed 25.8 TWh_e – some 30 per cent of electricity consumption in Finland (Statistics Finland 2005).

Another example is electric heating, which has been emphasized in the state power company IVO (now Fortum) campaign since the 1970s to create demand for nuclear power. It resulted in a 1,600 per cent increase in electricity consumption for heating, from 0.6 TWh_e in 1970 to 9.8 TWh_e in 2004, while over the same period total energy consumption for heating increased by only 1 per cent (Statistics Finland 2005). Coincidentally, over the same period the use of renewable energy for heating decreased by 26 per cent. Currently,

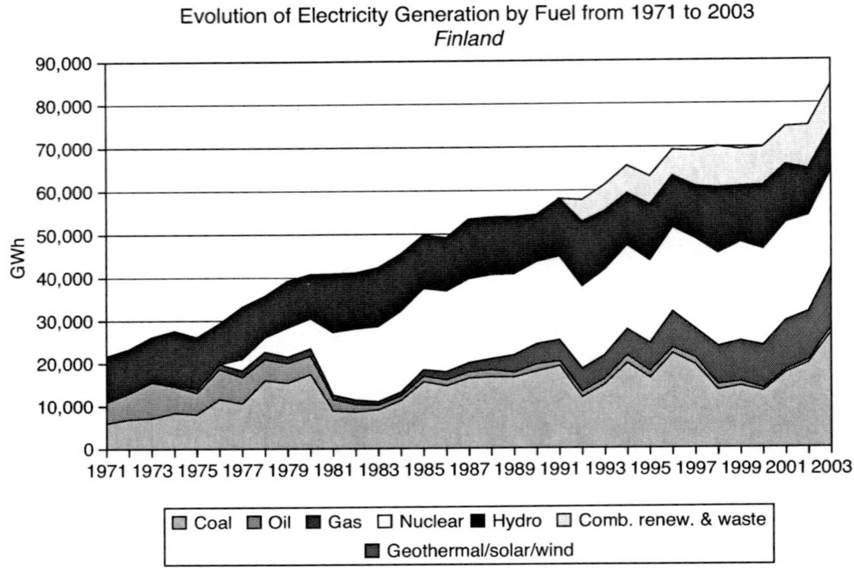

Figure 2.3 Electricity production in Finland increased by 280 per cent between 1971 and 2003. This is much faster than the global growth rate and twice as fast as in the OECD countries on average (IEA 2006)

direct electric heating has an 11 per cent share of Finnish electricity consumption and a 15 per cent share of heating consumption.

On average, since the first oil crises in the 1970s electricity production in Finland has increased twice as rapidly as in other industrialized countries (Figure 2.3). Over the period 1971–2003 the Finnish increase was 280 per cent, compared to 130 per cent in the EU-15, 150 per cent in the OECD countries and 210 per cent globally (IEA 2006). This growth is similar to that experienced in Latin America, suggesting that it is perhaps more characteristic of developing than industrial countries. Consumption has increased by an even higher amount (350 per cent) than production due to a substantial growth in net power imports. Over the same time period the share of electricity from renewable energy sources dropped from 47 per cent to 24 per cent (Statistics Finland 2005).

Thus, energy policy in Finland since the 1970s has not emphasized the domestically abundant renewable energy sources or the promotion of energy efficiency, with the exception of black liquor-based CHP production in the pulp industry. Lampinen (2000) has analysed some of the reasons for this development. At its core is the extraordinarily successful lobbying performance of heavy industry and the power industry, including many companies with a majority or otherwise significant state (in practice MTI) ownership, resulting in very low tax levels for large consumers and even the removal of CO_2 emission-based electricity tax in 1996 (Vehmas 2002). Finland is the only country in the world where CO_2 tax on power production has been removed (it remains in the heating sector).

The cold climate increases the level of heat consumption but, on the other hand, it makes CHP more economical, increases the efficiency of heat engines (like steam turbines in nuclear power plants), decreases cooling energy needs in heavy industry and increases the efficiency of wind power, solar power and solar heating. However, a peculiar domestic myth of the superiority of environmental performance of Finnish technology and the politically assumed difficulties in reducing emissions due to historical achievements compared to other countries has resulted in a free-riding policy at the UN and EU level despite the fact that international statistics do not support the myth (Lampinen 2004). Finnish aerial emission control targets within United Nations treaties on climate change (UNFCCC and its Kyoto Protocol) and long-range transboundary air pollution (UNECE CLRTAP and its Gothenburg Protocol) are among the weakest of any EU countries. And, according to the UN statistics based on the fourth national communications of the Rio Climate Convention, Finnish performance in limiting domestic greenhouse gas emission (+21.5 per cent compared to −1.4 per cent for the EU-15) is rated number 32 of 40 industrial countries (UNFCCC 2005).

The Finnish climate strategy project based on the UN Kyoto Protocol and the EU climate strategy has been seen by many as an opportunity to change the direction of domestic energy policy. Resorting to nuclear power could then be characterized as a missed opportunity, although nuclear power could also be used, together with renewables and energy conservation, to reduce the level of emissions.

Otherwise the climate change argument is the strongest of the arguments in the TVO (2000) application. There are no usage time greenhouse gas emissions from nuclear power and its lifecycle emissions are also low (UN 2000): not as low as those of wind power but lower than bioenergy, with the exception of the production of biogas from waste. Other aerial emissions from nuclear power are also very low. Therefore, nuclear power has a large short- and medium-term potential to decrease the level of greenhouse gas emissions if it replaces fossil and peat power production. However, no measures have been taken in Finland to reduce fossil power production and a feed-in tariff has been created to support peat energy. Finland is the only country in the world where feed-in tariffs are used to promote non-renewable energy and, curiously, *only* non-renewable energy. Renewable energy forms are not supported by feed-in tariffs, although there are plans to use this instrument for supporting biogas and wind – but still no other renewables.

In relation to the quantity of solid waste, nuclear power compares well with both fossil fuels and bioelectricity; however, nuclear waste, by its very nature, is extremely hazardous and causes an intergenerational environmental, health and military safety problem. For this reason and as a result of the low resource base of ^{235}U (OECD 2000), conventional fission power is considered by many not to be sustainable in the long run. As a consequence, nuclear energy is not mentioned in Article 2 of the Kyoto Protocol, where the means agreed by the governments for achieving the Kyoto Protocol targets are listed – including increased use of renewable energy forms and enhancement of energy efficiency. Nuclear energy has also been explicitly excluded from certain climate change policy mechanisms of the United Nations and the European Union. At the level of the UN, this applies to all industrial countries that have ratified the UNFCCC Kyoto Protocol, via the JI (joint implementation) project mechanism between industrial countries and the CDM (clean development mechanism) project mechanism between industrial and developing countries. At the level of the EU, this applies to all member countries and companies of the member countries via the ETS (emission trading scheme).

The UN-level restrictions on the use of nuclear power were set in the Marrakech Accords of the Kyoto Protocol which apply to all industrial countries, more specifically the UNFCCC Annex I countries that have ratified or will ratify the Kyoto Protocol (decided in the UN Convention in Marrakech in 2001). The JI restriction was made by decision 16/CP.7 titled 'Guidelines for the Implementation of Article 6 of the Kyoto Protocol': 'Recognizing that Parties included in Annex I to the Convention are to refrain from using emission reduction units generated from nuclear facilities to meet their commitments under Article 3, paragraph 1,...'. The CDM restriction was made by decision 17/CP.7 entitled 'Modalities and Procedures for a Clean Development Mechanism as Defined in Article 12 of the Kyoto Protocol': 'Recognizing that Parties included in Annex I are to refrain from using certified emission reductions generated from nuclear facilities to meet their commitments under Article 3, paragraph 1,...'.

Nuclear power has been excluded from the EU greenhouse gas emission trading scheme (ETS) by article 1 of Directive 2004/101/EC (EC 2004) amending Directive 2003/87/EC:

> ...in accordance with the UNFCCC and the Kyoto Protocol and subsequent decisions adopted thereunder, Member States are to refrain from using CERs and ERUs generated from nuclear facilities to meet their commitments pursuant to Article 3(1) of the Kyoto Protocol and in accordance with Decision 2002/358/EC, operators are to refrain from using CERs and ERUs generated from such facilities in the Community scheme...

Here CERs (certified emission reductions) are emission reduction certificates originating from CDM projects in developing countries and ERUs (emission reduction units) are emission reduction certificates originating from JI projects in industrial countries, both EU Member States and other industrial countries. Neither Member States nor companies are allowed to utilize nuclear power within the ETS. However, Member States as well as all other industrial countries can utilize nuclear power domestically in order to meet their Kyoto obligations. Within the JF (joint fulfilment) mechanism of the Kyoto Protocol that the EU utilizes by the so-called burden-sharing agreements, the status of nuclear power is unclear. In the original burden sharing agreement made in 1998 for the then 15 Member States, Sweden received a higher emission allowance (+4 per cent of the 1990 level by the average of 2008–12) since they promised to shut down several nuclear reactors. Since the shutdown rate was later slowed down Sweden unilaterally downgraded its emission allowance from +4 per cent to −4 per cent in the national climate law in 2001. If the same logic is applied to Member States that later decide to build new nuclear reactors, it would result in a tightening of the emission allowances by the amount of greenhouse gases that is theoretically replaced by the nuclear reactor. To date Finland is the only such country, but the Finnish government has no intention of making a unilateral downgrading of emission allowances similar to those made by the Swedish government. The burden-sharing agreement does not include modalities for the EU Commission to change the allowances of a Member State. The issue of dealing with both the increase and the decrease in the number of nuclear reactors will arise when the next burden-sharing agreement is negotiated for the 2nd Kyoto Protocol commitment period from 2013 onwards. On the UN level these negotiations are scheduled to be finished at the UN Climate Convention in Copenhagen in December 2009, allowing the EU burden-sharing decisions to be made in 2010.

Further separation between nuclear energy and renewable energies is established in the new EU Reform Treaty (Treaty of Lisbon) agreed to in October 2007. This amends the primary (constitutional) legislation of the EU by adding renewable energy as the only energy source that the EU energy policy should actively promote:

> 1. In the context of the establishment and functioning of the internal market and with regard for the need to preserve and improve the environment, Union policy on energy shall aim, in a spirit of solidarity between Member States, to:
>
>   (a) ensure the functioning of the energy market;
>   (b) ensure security of energy supply in the Union; and
>   (c) promote energy efficiency and energy saving and the development of new and renewable forms of energy;
>   (d) promote the interconnection of energy networks.

In addition, climate change is added by the Treaty of Lisbon as the only explicitly mentioned environmental problem in the goals of EU environmental policy. In the energy and climate policy package released in January 2008 by the European Commission, binding targets for the increase in renewable energy are set for every Member State, but no goals are set for nuclear power.

Thus, in relation to the mitigation of climate change both the United Nations and the European Union give renewable energy preference over nuclear fission power. Nuclear fusion power is a possible future technology, which utilizes more abundant resources and generates much less toxic radioactive wastes with a much shorter lifetime (UN 2000). However, it is completely different and speculative technology that is not relevant to the considerations of the fifth reactor project.

Many renewable energy sources and energy conservation technologies offer lower or non-existent lifecycle emissions of greenhouse gases, other aerial emissions and solid waste as well as a large resource base (UN 2000). The direct environmental merits of nuclear power compared to fossil power are great and, compared to wood burning, significant. But the merits are negative compared to those of wind power, biogasification and many other renewable power sources as well as energy conservation, all of which offer long-term sustainable solutions.

It is also necessary to take into account the climate and other ecological effects of additional load balancing power required by the fifth reactor and the other potentially forthcoming nuclear reactors that cannot be adjusted according to consumer needs. The balancing power options include fossil power and hydropower. Several influential politicians, in particular minister of trade and industry Mauri Pekkarinen, and the energy industry have demanded increasing hydropower capacity through the removal of the barriers created by current environmental legislation, like the Water Act (264/1961), the Rapids Protection Act (35/1987) and special laws protecting individual rivers, as well as the case law produced by the Finnish Supreme Administrative Court. The main source would be a new Vuotos reservoir in Northern Finland, as was proposed by the state company Kemijoki Ltd in 1974. The Supreme Administrative Court judgement in 2002 (KHO:2002:86) denied building permission for the Vuotos reservoir on the basis of widespread negative environmental impacts of the reservoir. The decision is final, meaning that the only way to reopen the project is to change the law on which the decision was based, that is section 2:5 of the Water Act. Currently a modification to the Act is being prepared to enable government to deny on a case-by-case basis the application of section 2:5 in hydropower plant permitting. The reason given to the public for reopening the Vuotos case is not the need to add load balancing power for the new nuclear reactors, but, rather, mitigation of climate change by increasing emission-free hydropower production. However, the peak greenhouse gas emissions of the reservoir are 4 Mt CO_2^{eqv} for 0.36 TWh of additional annual power production (Väisänen et al. 1996). This is 14 times higher than the greenhouse gas emissions of condensing coal power plants per electricity unit produced, assuming a 100-year perspective in the global warming potentials (GWP) of methane, carbon dioxide and nitrous oxide. Since the majority of the emissions are methane, as a result of anaerobic decomposition of a large amount of biomass under the shallow water of the reservoir, the emissions are further threefold higher when the more relevant 20-year GWP perspective is used. Although the reservoir emissions will diminish substantially five years after the inundation, the level of greenhouse gas emissions is large compared to fossil power and therefore there should be no factual grounds for claiming climate change mitigation merits for the project. However, the project is very useful as a balancing power source for nuclear reactors and it is also known to be an economically attractive investment for the prospective owner Kemijoki Ltd, a company in which the state holds over half of the shares and Fortum has 17.5 per cent. On the other hand, PVO has already begun an EIA process on building the Kollaja reservoir in Northern Finland for balancing nuclear power. Kollaja, like many other potential hydropower plant options, is protected by the Rapids Protection Act and several politicians have suggested a release of those sites by amending the act. In the future, pumping storage reservoirs will also be proposed. To date only one application (at Vaarunvuori in Central Finland) has been made, but in 2000 the site became protected under the EU Habitats and Wild Birds Directives following a judgement of the Supreme Administrative Court (KHO:2000:40). Removing current nature protection from potential hydropower reservoir sites in order to build balancing power for NPPs will have diverse and widespread effects in nature, including substantial greenhouse gas emissions. Those emissions are not under current legislature taken into account in the permitting process for new plants or the greenhouse gas balance of operating companies. Those emissions are also ignored in the EU emission trading system and national greenhouse gas balances.

Another issue relates to acute catastrophic phenomena as risks inherent in nuclear power technology. Nuclear accidents usually have only a local impact, and at most a regional impact, as was the case with Chernobyl. They do not have an impact on global environmental problems. However, nuclear power technology enables nuclear and other radioactive weapon technology that does have great potential for serious environmental problems locally, regionally and globally.

An additional argument given in several statements and parliamentary discussions was the impact on developing countries – some of which, such as Iran, have their own uranium resources – if the industrial countries were to slow down the market growth of renewable energy technologies in favour of nuclear power. Both fossil and nuclear energy would benefit, raising both environmental and security risks on a global level. The Finnish decision could trigger this development since it is cited as an example in many countries (Underhill 2006).

Thus, the possibility of supplying the electricity market with nuclear power and hampering the development of domestic renewable energy production and industry is an important issue. This issue was addressed by the Finnish Parliament as a renewable energy promotion package requirement as a condition for its positive decision on the TVO application. To date this policy requirement for coexisting nuclear and renewable energy promotion has been implemented only poorly and the MTI proposal for a new national climate and energy strategy submitted to Parliament in November 2005 did not seriously consider the implementation of that requirement in either the short or the long term. Nevertheless, it was approved by Parliament.

Energy security benefits: Security of electricity production and energy import dependency

The security of energy supply argument advanced in the TVO (2000) application was based on one hand on the availability of uranium from many politically stable countries, and on the other on the ability, due to the high energy density of the fuel, to store large reserve stocks, in practice currently for two years' consumption. But these valid arguments were not balanced by several aspects that decrease the level of energy security.

It was not mentioned that the proven global uranium reserves, assuming at most four times the current price, are adequate for only 50 years at the current rates of global consumption (OECD 2000), which is less than the assumed 60-year lifetime of the fifth reactor. Instead, the application includes an embedded technology optimism presuming that it will always be possible to acquire new cheap uranium on the market as and when necessary. On the other hand, the application includes embedded market pessimism concerning renewable energy sources that have very large proven reserves (UN 2000).

Furthermore, it was not mentioned that the security of energy supply may be decreased by:

- increased reliance on imported fuels;
- decreased diversity of electricity supply; and
- strengthening of the centralization trend of electricity production.

Currently, the mix of Finnish electricity generation is very diverse, but nuclear power is already the largest component: 28 per cent in 2006. Thus, the addition of more reactors will result in the domination of the mix by nuclear electricity and a high dependency upon it, as depicted in Figure 2.4. Still, TVO (2000) advances the argument that additional nuclear power will improve the level of Finland's energy diversity.

The Finnish level of self-sufficiency in primary energy (31 per cent in 2004) and electricity supply (37 per cent in 2004), including a significant contribution from non-renewable peat (Statistics Finland 2005), are critically low,

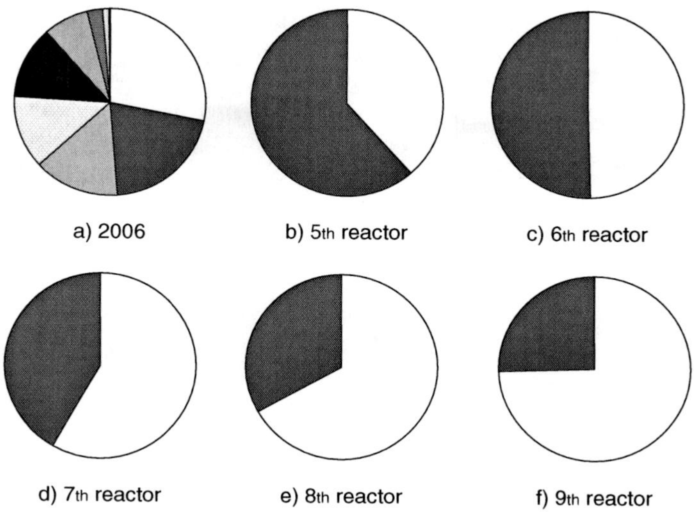

Figure 2.4 Decreasing diversity of electricity generation by adding nuclear reactors (blank slice) to the Finnish mix in 2006

causing a high degree of vulnerability in case of an energy crisis. The construction of new nuclear reactors contributes to this vulnerability since Finland has no domestic reserves of uranium, nor fuel production facilities or a reactor industry. All these need to be imported, as along with the construction know-how and much of the workforce needed to carry out the building work.

Decreasing the level of power imports from Russia, Sweden and Norway would increase the level of energy security, if it were to be the result of adding new domestic generation capacity. However, in practice Finnish industry will continue to import as much power as it can since old nuclear power in Russia and old hydropower from Sweden and Norway are cheaper than any new power that might be generated and also cheaper than most of the existing power produced in Finland. Thus, the new reactor is not likely to decrease the level of power imports. On the contrary, large new import cable projects have been proposed. Power imports may be decreased, however, as has already occasionally happened and is more likely to happen in the future, due to increased peak power demand in countries exporting to Finland coincidentally with peak power demand in Finland.

In recent years blackouts and high price peaks (even above €1/kWh) have become increasingly common in Finland (EMV 2006a). According to the Energiakolmio consulting company, the price peaks are causing €40–400 million of additional annual costs for Finnish electricity consumers. The blackout costs for electricity producers were €5.6 million in 2005, a figure that would be much higher if full compensation to consumers were required, such as might occur in the future as the result of pressure from the EU. One of the main reasons for these problems with the Finnish power grid is the centralizing trend of electricity production that is being strengthened by the fifth nuclear reactor and the current plans for additional reactors. These problems are likely to become even worse due to an increase in extreme weather events, the decreasing stability of international energy markets, a decrease in the level of both grid maintenance and consumer protection in the liberalized markets, increasing peak demand and an increasing threat of terrorism and vandalism. The fifth reactor will be responsible for 10–15 per cent of Finnish power generation. Losing such a high share of generation, either unintentionally or intentionally, would lead to severe large-scale blackouts, as experienced in recent years, for example, in the United States (both intentional and unintentional) and Italy (unintentional). This vulnerability is amplified by having two other nuclear reactors and another 10–15 per cent of power generation at the same location in Olkiluoto and the main power cable connection to Sweden, with another 10–15 per cent potential, in the immediate vicinity. The case of building the fifth reactor in Loviisa would have been rather similar, since there are two existing reactors on site and the main Russian import power cable runs nearby.

Thus, this level of centralization would need a level of corresponding reserve capacity. Additional capacity will also be required for load balancing, since the nuclear power generation profile is inflexible, with no capability to respond to normal rapid fluctuations in consumption. This may increase greenhouse gas emissions as described earlier, if hydropower or fossil power is used for the purpose, and it may decrease the potential of intermittent renewable power sources, especially wind power, since they may need to compete for the balancing power with nuclear reactors. The large additional power capacity requires a strengthening of the grid and increases the pressure to build electricity storage facilities, using pumping storage, compressed air storage or other technologies. Currently, these technologies are not being used in the Finnish grid. The TVO (2000) application did not address the economic implications of reserve and balancing power capacity, grid strengthening and storage. Part of this problem is addressed by a 100 MW_e gas turbine power plant built at Olkiluoto by Fingrid. This turbine offers the national grid some security of supply in the case of shutdowns. But, more importantly, it offers security for the internal electricity needs of the three nuclear reactors in the case of emergencies. In October 2008 Fingrid started an EIA process for building four additional large reserve power plants distributed in various parts of Finland. The costs of reserve power plants are not added to the price of nuclear electricity specifically, but will be paid for by every power consumer.

Many statements pointed out that these energy security concerns would decrease through the use of decentralized renewable energy technologies. The argument is that a large number of small electricity producers would create immunity against most blackouts caused by transmission, distribution and centralized power plants and that the problem of the dependency on imports would vanish by utilizing abundant local and national renewable energy resources. The small producers could also increase the quality of grid electricity by using frequency converters and decreasing the need for costly grid strengthening by adding local production in the same areas where consumption increases. Furthermore, in a crisis situations, such as caused by terrorism, vandalism, war or market forces, it would be impossible to paralyse the whole power system if it was based on a large number of small producers, whereas the present Finnish power system could be shut down by targeting just five or ten of the most crucial vulnerabilities. For example, in military doctrines, targeting the vulnerabilities of a power system is among the highest priority actions and always takes place in the first hours of conflicts. Targeting NPPs may have added value as they may become radioactive weapons.

Opposing statements were also given. The most noteworthy of these was the statement by the Ministry of Defence. This implied that in crisis conditions it is easier to defend a few centralized power plants rather than many distributed ones, that it is better to rely on imported uranium instead of domestic energy sources, and that possible attacks on NPPs are not considered to be a threat.

Economic benefits: Competitive and stable price of nuclear electricity

The competitive price argument in favour of nuclear power was based on a comparison with the costs of producing condensing coal, natural gas and peat power. The comparison study was made at Lappeenranta University of Technology as an MSc thesis project (Rissanen 1999) that was commissioned by TVO and used in the TVO (2000) application. This showed that nuclear power could be generated at 2.2 cents/kWh, whereas the costs of coal, natural gas and peat were calculated at 2.4, 2.6 and 3.1 cents/kWh respectively. This price for nuclear electricity in Finland is lower than in any other country, including developing countries and those countries with own fuel resources, own fuel production industry and own reactor technology (OECD 1998; Lampinen 2002). Although the most up-to-date OECD economics study at the time (OECD 1998) was used by TVO (2000) in other contexts, it was not mentioned that the OECD estimate for new Finnish nuclear electricity was more than 40 per cent higher than the TVO estimate (Figure 2.5). The TVO estimate was criticized in many expert statements. It was regarded as being technologically optimistic in respect of nuclear power and technologically pessimistic in respect of renewable energy sources.

Furthermore, there was no mention of other power generation technologies, such as wood cogeneration, which is highly competitive in Finland (Figure 2.5). But, according to Fortum (1999), wind power was presumed to reach competitiveness by the start-up time of the fifth reactor. One wind turbine was erected by TVO at Olkiluoto to gain experience of that technology.

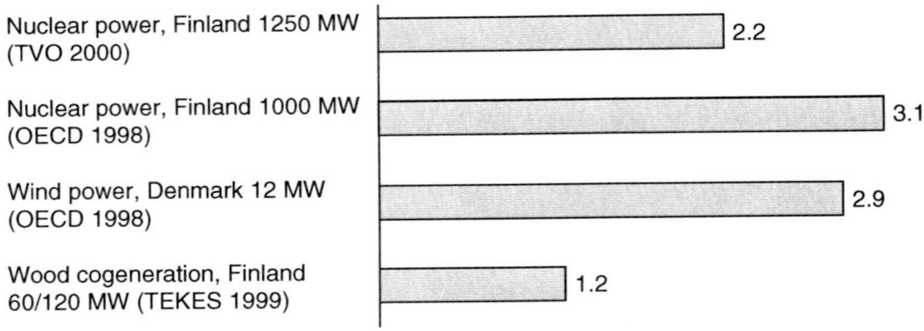

Figure 2.5 TVO and other electricity price estimates in eurocents/kWh (Lampinen 2002)

There are many distributed renewable energy technologies that are not competitive with nuclear power in the current Finnish market conditions, as a result of both attitudes and economics. The situation will remain so until a conscious political choice is made to promote such technologies.

There are some reasons why new nuclear power could be cheaper in Finland than in other countries, in addition to the high Capacity Factor caused by high production efficiency inherent in a cold climate with few production interruptions. The chosen reactor is the first sold EPR reactor and acts as a technology demonstration, carrying the usual risks for the purchaser. It was also the first reactor to be ordered in Western Europe since 1986 and it has therefore been considered crucial for the revival of the nuclear power industry in Europe and beyond (Underhill 2006). These policy related factors resulted in a low bidding price of €3 billion.

Low-interest financing was received from a bank group under the leadership of the German bank BLB and other sources, export guarantees were given by the French national export credit agency and long-term power purchasing agreements with municipal and private companies were also signed (EREF 2004). The European Commission investigated whether or not these financing arrangements, which include the public participation of several Member States and municipalities, are consistent with the regulations concerning state aid and public procurement (EREF 2004). In September 2007 the Commission ruled that the regulations had not been violated by the French national export credit agency.

The investment cost estimate of the reactor given by TVO (2000) was between €1.7 and 2.5 billion. However, the minimum estimate increased to €3 billion as a result of the bidding process and estimates of €3.2 billion were given as early as 2003 by the credit rating agency Standard and Poor's (EREF 2004). Since then the cost has risen due to several errors made in the construction, causing at least three years of delay on the planned reactor start-up time in 2009, as given in the TVO (2000) application. On 10 September

2007 a press release by ElFi, a power market company of Finnish electricity consumers, estimated the cost of this delay at €3 billion. In January 2009 the Finnish Broadcasting Company, based on a Siemens quarterly report, reported that TVO has made a claim of €2.4 billion from constructors Areva and Siemens. On the other hand, in December 2008 Areva and Siemens demanded a €1 billion delay compensation from TVO.

Since the new reactor was seen in the Finnish climate strategy (MTI 2001) as being the primary way of achieving the UN Kyoto Protocol and the EU burden-sharing commitments in 2008–12, the delay will make it impossible to reach the emission limitation target.

According to TVO (2000), half of the investment was meant to be spent on domestic work, materials and equipment, but in October 2007 TVO admitted that this goal will not be achieved. In any case the project will be an important employer. However, the level of employment would have been higher if similar electricity production capacity had been built using domestic renewable energy sources. One benefit of renewables, as quoted by the European Commission, is that distributed renewable energy production has approximately five times the employment effect compared to centralized fossil and nuclear power production per each kWh of electricity produced (EC 1996). Renewables employ mostly less educated people in rural areas nationwide, whereas nuclear power employs highly educated people locally.

The cost of reserve and balancing power, grid strengthening and energy storage was not included in the TVO cost estimate or in the statements of Fingrid to the MTI and the MTI to government in the case of either the Olkiluoto or the Loviisa site for the fifth reactor. However, the proposed United Power 1,000 MW_e cable from near Sosnovyi Bor to near Loviisa would mean more than €2 billion costs to the Finnish grid according to Fingrid and over €1.2 billion according to Finnish Energy Market Authority EMV (EMV 2006b). These costs arise from investment needs for grid strengthening within Finland and to Sweden, and for reserve power plants. There is no reason to believe that the fifth reactor would not create corresponding needs and costs. It was just not politically convenient for Fingrid to take them up in the DiP process. The MTI has 50 per cent voting rights in Fingrid and most of the rest is with the industry that applied for the fifth reactor. On the other hand, the independent United Power cable would have imported power for all power consumers, rather than just heavy industry. MTI decided not to grant permission for that cable.

The flat profile of nuclear power production is ideal for satisfying the base load, if the base load is defined as flat, as is customarily the case. However, the actual load curve in Finland shows very strong seasonal variations as the result of the northern climate: the base load in wintertime is much higher than it is in summertime. In Finland, the annual production curve of wind power and biopower (CHP) is similar to the load curve (Lampinen 2002). This means that the more efficient use of these technologies, together with current power production, could decrease the total cost of the electricity system by decreasing the need for reserve power.

One issue that could not be included in the TVO (2000) application was an amendment written on 25 June 2005 in the Finnish Nuclear Liability Act (484/1972). This increased the accident liability of the NPP operator from €185 million to €700 million – a change that was the minimum allowable by the 2004 amendment to the OECD Paris Convention of 1960 on Third Party Liability in the Field of Nuclear Energy. The convention parties have the liberty to set the limit above €700 million, but in Finland this was not done. The reason given by MTI was that in Finland, unlike other countries, it would be difficult to get insurance for higher liability. This liability needs to be covered by insurance and will add to the price of nuclear electricity. The Finnish government takes responsibility for damages exceeding €700 million. As a comparison, the cost of the Chernobyl accident has been put at more than €100,000 million. Thus, the purpose of the Liability Act is to remove the operator's financial responsibilities in large-scale nuclear accidents. This has been a prerequisite for the viability of the commercial nuclear power industry (Roggenkamp et al. 2001) since the pioneering Price–Anderson Nuclear Industries Indemnity Act, which was enacted in the USA in 1957. The new amendment accepted by the Finnish Parliament in 2005 is still not in force and pressures for changes currently exist to remove all liabilities regarding environmental damage, cancers and other health impacts with long incubation period, and also any damage caused by terrorism. Under the proposal, the responsibilities for covering such impacts would be moved to the government and private citizens. Therefore, if the nuclear liability act is so modified, the insurance costs would decrease and the price of nuclear power would also decrease.

One benefit of nuclear power, compared to both fossil and also biopower, is that uranium fuel is cheap, being currently only about 15 per cent of the price of nuclear electricity (TVO 2000). Market price variations of the scale experienced in crude oil and natural gas markets would have little effect on the price of nuclear electricity, but the situation might well have changed by the end of the 60 years' planned lifetime of the fifth reactor. According to reports from the UN International Atomic Energy Agency (IAEA) and the OECD Nuclear Energy Agency (NEA), at the rate of current consumption the known economic reserves of ^{235}U will last for around 50 years (OECD 2000). Here the economic limit is at most four times the current cost. In addition, the ^{235}U in existing nuclear weapons would be enough for a decade. Using that fuel source would be desirable for security reasons, not only for reducing nuclear weapons, but also for reducing depleted uranium weapons. According to the Canadian Cameco company, the world's leading uranium producer, the price of raw uranium has already risen tenfold since the TVO (2000) application was made (Copeland 2007). This has been driven principally by the expansion of nuclear power programmes in both China and India. However,

TVO (2000) did not anticipate a significant price rise of processed uranium fuel during the lifetime of the reactor.

Discussion

Unlike other types of power plants, nuclear reactor project applicants are not held responsible under the Nuclear Energy Act for the publication of inaccurate or false application data. The main incentive for the applicants of nuclear reactors is to create a politically convenient application to sell the idea; data accuracy is of lesser importance. The political Decision-in-Principle (DiP) process replaces the technical environmental permit process of other types of power plants. The DiP process also provides legitimation for all political opinions in the government and in Parliament.

An analysis of the arguments presented in the DiP application for the fifth reactor indicates that some of them are valid, whereas some are unlikely to be correct and some have already been shown to be false. As an example of the latter, the current construction cost estimates are at least 100 per cent higher than the upper limit of the cost range given in the application; the current estimate for reactor construction time is at least 70 per cent more than the estimate given in the application, with significant cost implications; and the target of at least a 50 per cent share for domestic investment will not be achieved. Both the low cost and the rapid building time, with its implications for fulfilling the Kyoto Protocol targets, were major arguments for receiving support from members of Parliament.

Although the review of the application highlighted numerous suspect arguments from many experts, related to environmental, employment, security of energy supply and economics, only the security shortcomings mentioned by STUK were corrected, because of the legal authority STUK has in this matter. The MTI policy as a responsible authority in the DiP process was similar to the MTI policy as a responsible authority in the waste disposal EIA procedure, where numerous expert arguments given in the statement were omitted from the MTI statement and also from the overviews of statements (Hokkanen and Ruuskanen 2005). The situation was different in the parliamentary expert review process. Hylkilä (2003) found it to be a high-quality process in which a lot of technical information was conveyed and many members of Parliament considered it important for their decision, although lobbying within Parliament was clearly dominated by arguments in favour of the reactor.

The Finnish Nuclear Energy Act offers, in principle, a very broad and democratic decision-making framework. But in practice power is shared between only a few civil servants at the Energy department of MTI who are the responsible authorities in EIA, DiP, building permit and utilization permit processes; the major owners of applicant companies and other relevant companies; the supervisors of relevant national authorities; and the formulators and implementers of Finnish energy and climate policy, research and funding. One result of this concentration of power has been the ability of MTI to prevent external expert arguments of choice from being given serious consideration. No resources were allocated for the independent evaluation of criticism presented in the expert statements and no policies were implemented for dealing with them. The MTI established a public–private partnership success story with the applicant companies, in which the MTI also had considerable influence as a major shareholder on behalf of the Finnish government. Consequently, the MTI can guarantee the smooth handling of nuclear facility applications, despite the criticism presented in the review process.

During the first half of 2010 the Finish government is expected to make decisions about three new DiP applications. Learning that some of the arguments used in the DiP application to secure support from ministers and members of Parliament in the case of Olkiluoto-3 turn out to be incorrect would presumably lead to some policy changes when dealing with the forthcoming reactor applications. This could have an influence on energy policy and the use of expert knowledge in policy processes.

Notes

1. Furthermore, environmental impact assessment (EIA) needs to be made prior to the application according to the Finnish EIA Act (468/1994) and EIA Decree (268/1999). The MTI is the responsible authority for EIA processes concerning nuclear energy installations. (See Appendix 1 in Chapter 8.)

References

CEPI (2002) *Energy Profile 2001*. Brussels: Confederation of European Paper Industries.

Copeland, M.V. (2007) *Dogging the Nuclear Future*. Business 2.0, August 2007, 84–9.

EC (1996) *Energy for the Future: Renewable Sources of Energy, Green Paper for a Community Strategy*. COM(96)576, 20 November 1996, Brussels: European Commission.

EC (2004) Directive 2004/101/EC of the European Parliament and of the Council of 27 October 2004 amending Directive 2003/87/EC establishing a scheme for greenhouse gas emission allowance trading within the Community, in respect of the Kyoto Protocol's project mechanisms. *Official Journal of the European Union* L 338 (13.11.2004), 18–23.

EMV (2006a) *Annual Report 2006 to the European Commission*. Helsinki: Energy Market Authority, Finland.

EMV (2006b) *Statement of EMV for the Ministry of Trade and Industry 28/8/06* (in Finnish), Helsinki: Energy Market Authority, Finland.

EREF (2004) *Complaint for the European Commission Concerning the Purchase by TVO of the Framatome ANP 1600 MW Nuclear EPR Reactor to a Fixed Price and Under Various Support Schemes from Member States*. European Renewable Energies Federation & Kuhbier law firm, Brussels 14 December.

Fortum (1999) *Loviisa 3 -ydinvoimalaitoshanke – Ympäristövaikutusten arviointiselostus* (EIA report of Loviisa-3 nuclear reactor). Vantaa: Fortum Power and Heat Oy.

Hokkanen, P. (2004) *Kansalaiskeskustelu lisäydinvoimasta – mielipidekirjoitukset osallistumiskanavana* (Public discussion on additional nuclear power capacity – Written public opinions as a channel of participation). MTI Financed studies 6/2004, Helsinki: Ministry of Trade and Industry.

Hokkanen, P. and Ruuskanen, T. (2005) *Kirjallisten mielipiteiden vaikuttavuus YVA-menettelyssä* (Impact of written statements in EIA processes). Alueelliset ympäristöjulkaisut 396, Keski-Suomen ympäristökeskus, Jyväskylä.

Hylkilä, I. (2003) *Suomen viidennen ydinvoimalan lobbaus kansanedustajan näkökulmasta* (Lobbying process in the Finnish Parliament regarding the 5th Finnish nuclear reactor). MA thesis in Communication, University of Jyväskylä.

IEA (2006) IEA Energy Statistics Online. OECD International Energy Agency, Paris, www pages: www.iea.org/Textbase/stats/index.asp (15/9).

IPCC (2000) *Special Report on Emission Scenarios*. UN Intergovernmental Panel on Climate Change, Cambridge: Cambridge University Press.

Kautto, N. (2005) *Analysis of Policy Options and Implementation Measures Promoting Electricity from Renewable Biomass in the European Union*. DG JRC, European Commission.

Lampinen, A. (2000) Phenomena in Finnish Climate Economics – How Will Energy Efficiency and Other Win–Win Opportunities be Taken into Account in the Finnish National Climate Change Strategy? (in Finnish) *Futura* 19(3), 81–95.

Lampinen, A. (2002) Huomioita kansantaloudellisten kustannusten käsittelystä Suomen energiatalouspolitiikassa (Observations on handling national economic costs in Finnish energy economics policy). *The Finnish Economic Journal* 98, 207–19.

Lampinen, A. (2004) Teknologian suuri kertomus Suomen ilmastopolitiikassa (Myths of technology in Finnish climate policy). *Kanava* 32(2), 96–9.

MTI (2001) *National Climate Strategy – Government Report to Parliament*. Publication 5/2001, Helsinki: Ministry of Trade and Industry.

OECD (1998) *Projected Costs of Generating Electricity: Update 1998*. Paris: OECD Nuclear Energy Agency and OECD International Energy Agency.

OECD (2000) *Uranium 1999: Resources, Production and Demand*. Paris: OECD Nuclear Energy Agency and UN International Atomic Energy Agency.

Rissanen, S. (1999) *Ydinvoiman kilpailukyky ja merkitys Suomen energiataloudessa* (Competitiveness of nuclear power and its contribution to the Finnish energy economy). MSc thesis in Engineering, Lappeenranta University of Technology.

Roggenkamp, M.M., A. Rønne, C. Redgwell and I. del Guayo (eds) (2001) *Energy Law in Europe – National, EU and International Law and Institutions*. Oxford: Oxford University Press.

Statistics Finland (2005) *Energy Statistics 2004*. Energy 2005:2, Helsinki: Statistics Finland.

TEKES (1999) *Bioenergy Research Programme Final Report 1993–1998*. Bioenergy Research Programme publications 30, Jyväskylä: TEKES and Jyväskylä Science Park Ltd.

Tirkkonen, J. (2000) *Ilmastopolitiikka ja ekologinen modernisaatio* (Climate policy and ecological modernisation). Acta Universitatis Tamperensis 781, Tampere: University of Tampere.

TVO (1999) *Olkiluodon ydinvoimalaitoksen laajentaminen kolmannella laitosyksiköllä – Ympäristövaikutusten arviointiselostus* (EIA report of Olkiluoto-3 nuclear reactor). Olkiluoto: Teollisuuden Voima Oy.

TVO (2000) *Ydinvoimalaitosyksikön rakentamista koskeva periaatepäätöshakemus – Loviisa 3/Olkiluoto 3* (Application for DiP for 5th Finnish nuclear reactor). Olkiluoto: Teollisuuden Voima Oy.

UN (2000) *World Energy Assessment – Energy and the Challenge of Sustainability*. United Nations Development Programme, New York: United Nations Department of Economic and Social Affairs and World Energy Council, UNDP.

Underhill, W. (2006) A Change in Climate – Political Worries are Driving Nuclear Rethink in the West. *Newsweek* 6 February, 37–8 (and related articles on pages 30–9).

UNFCCC (2005) *Key GHG Data – Greenhouse Gas Emissions Data for 1990–2003 Submitted to the United Nations Framework Convention on Climate Change*. Bonn: UNFCCC Secretariat.

Väisänen, T.S., P. Martikainen, S.K. Hellsten, A. Niskanen, J. Huttunen, M. Heiskanen and O. Nenonen (1996) *Importance of Greenhouse Gas Emissions from Finnish Hydropower Production Compared to Other Sources of Electricity, with the Kemijoki Water Body as an Example*. Montreal: IAEA Advisory Group Meeting on Assessment of Greenhouse Gas Emissions from the Full Energy Chain for Hydropower, Nuclear Power and Other Energy Sources.

Vehmas, J. (2002) *'Rahat Ruotsiin ja päästöt Tanskaan' – Suomen ympäristöperusteisen energiaverotuksen rekonstituutio 1993–96* ('Money for Sweden and CO_2 emissions to Denmark' – Reconstitution of the Finnish environment-based energy taxation in 1993–96). Acta Universitatis Tamperensis 861, Tampere: University of Tampere.

VN (2002) *Valtioneuvoston periaatepäätös 17 päivänä tammikuuta 2002 Teollisuuden Voima Oy:n hakemukseen ydinvoimalaitosyksikön rakentamisesta* (Government decision on the TVO application for DiP). M4/2001vp, Helsnki: Valtioneuvosto (Government).

Finnish nuclear power legislature

Laws (www.finlex.fi/en/):

- Nuclear Energy Act (990/1987)
- Nuclear Energy Decree (161/1988)
- Nuclear Liability Act (484/1972) (old name law on atomic liability was used until 1989)
- EIA Act (468/1994)
- EIA Decree (268/1999)

Other regulations

(www.stuk.fi/ydinturvallisuus/ydinvoimalaitokset/saannosto/en_GB/saannosto/)

3

Social Dynamics Behind the Changes in the NGO Anti-Nuclear Campaign, 1993–2002

Harri Lammi

Democratic pragmatism and the production of debate

To examine how national environmental questions are dealt with in Finnish society it is necessary to outline some background assumptions in the fabric of the decision-making processes. This chapter will consider the assumptions of democratic pragmatism through case studies of the nuclear decisions made in 1993 and 2002.[1] In short, this branch of thought has emphasized the participation of different social groups in the environmental decision-making process (Dryzek 1997). The impacts of this school of thought have included, for example, establishing a system of citizens' participation programmes in the Finnish environmental impact assessments and the involvement of non-government organizations (NGOs) in the work of parliamentary committees. One of the basic assumptions has been that public participation is the key to solving environmental problems and that the level of civil society participation should, therefore, be extensive. In theory, participation processes should give different stakeholders the possibility to present their cases either for or against certain propositions.

Considering the efforts and processes ensuring public participation, Finnish society has had relatively little public discussion around the 'production of debate' – for example, the materials production behind the environmental arguments advanced by different organizations and movements in society. This debate has been more important in scientific circles than in the Finnish political arena (for example, Litmanen 2001). Further research relating to the political usage of science is definitely required (see Litmanen 2008). Thus, although the Finnish parliamentarians seemed to be aware of the disparity of resources between organizations, the assumption behind the parliamentary hearing system in nuclear decision making seemed to be that the 'best argument wins'.

The debates of 1993 and 2002 surrounding the construction of a new NPP unit illustrate that what counts as a 'best' argument is obviously a very complex question. One could argue that the production of arguments and the 'neutral' research information behind these arguments has been central to the art of politics. In the era of cut-and-paste word processing, arguments are passed around quickly and are repeated in different forms and mixtures.

Thus, those organizations which control the production and reproduction or arguments have a great deal of power in society. What does the production of arguments then need in the cases we are talking about? The production of information about nuclear power was principally undertaken by a number of different groups, including the power companies, the Radiation and Nuclear Safety Authority (STUK), the Ministry of Trade and Industry, the Ministry of Environment, Technical Research Centre of Finland, and the NGOs. The situations and resources of these organizations are obviously very different.

Various argumentation and information material supporting and criticizing the new NPP unit project were reproduced and modified – and sometimes even silenced – by the media. Among examples of information production are the Lappeenranta Technical University reports about the low costs of a new nuclear reactor which were published before the nuclear debate in 2001 (Tarjanne and Rissanen 2000). These reports were widely quoted in the media and in the political arena. The production and effective dissemination of such material favourable to nuclear power requires the establishment of relatively well-funded research organizations.

If a certain amount of financial capital is important to the production and dissemination of information materials, social capital plays an even more important role in the dissemination of the arguments. Obviously, the provision of funding also affects the possibility of attracting credible researchers to produce and present technological or economical arguments, which gives the arguments a greater chance of being widely used as credible arguments.

A vivid case highlighting the importance of social credibility in certain fields of debate was the use of economic arguments against nuclear power by some Finnish environmental NGOs. Their cause was not helped by the fact that NGOs only made use of arguments produced earlier by well-established state organizations and universities. Journalists kept asking why NGOs were now interested in the economics of nuclear power, and expressed doubt that environmentalists had a better grasp of the economics than those companies willing to invest in nuclear power. Few actually examined the original materials which had been used by the NGOs. The authority of the NGO campaigners in economic spheres was not sufficient for the advancement of politically credible economic arguments.

The problems analysed below may allow one to ask critical questions of basic tenets of Finnish decision making in respect of the environment. Regardless of the imbalance in resources between, for example, NGOs, state organizations and companies, it is often taken for granted that NGOs or individual persons can produce sufficient materials and arguments to have meaningful and balanced debate about environmental decisions. Or, if they cannot, that this only proves that the other side was right in their argumentation. The author states that this kind of assumption is problematic, because it places a heavy burden on civic organizations, which are under-resourced to act as effective counter-debaters to company or state actors. The fact that NGOs have, in many cases, produced materials and campaigns which have had great significance for decision making has also been dependent on the commitment on individual people and organizations. The continuity of such activism cannot be relied on as a basis of environmental debate in the society.

The role of environmental and research institutions is interesting in this respect. Although researchers have participated in certain environmental debates one might claim that the participation of institutions in recent environmental debates, such as the nuclear debate in 2002 (see Hokkanen 2004), has not been particularly significant. It might be that these institutions are not asked to participate or interfere, or that when they do participate they do not dare to offer overly harsh criticism, with few notable exceptions, to which we will return later. It has even been the case in recent nuclear debates that researchers potentially critical to nuclear power in the state-funded 'objective' energy research organizations have been denied access to politicians.

For whatever reason, in recent debates the role of critic has often fallen to NGOs. The results are predictable, in the probabilistic sense of the word. This is not to claim that resource imbalances between actors are the sole determinant of the outcome of lobbying and campaigning. They are not. Often, it has been the case that small David has struck straight into the forehead of Goliath. But perhaps we have been so mesmerized by this symbolic, almost idealistic imagery of democratic environmental decision making that we fail to see the statistical fact: most of the time Goliath has won. Resources do not determine the outcome, but they often help.

Introduction to the Finnish nuclear decision, 1993–2002

The chapter is based on the author's forthcoming PhD.[2] The research material consists of several thousand pages of transcribed parliamentary debates during the DiP processes of 1993 and 2002, the debate on the government energy strategy of 1992, the DiP about nuclear waste and the debate on the government climate strategy of 2001. This is particularly rich material for understanding the storylines, as the majority of the parliamentarians played a very active part in the discussions. In addition to the main material mentioned above there are thousands of pages of related government proposals, committee statements, and lobby materials from the industry and the NGOs. The method applied is a textual analysis relying on the Foucaultian tradition with some novel analysis methods (compare Ylönen 1999; Sjölander 2004). The analysis about the formation of the NGO campaign relies on interviews with the major actors from the 1992–93 and 2000–02 campaigns and the recorded memories of the actors (Nissinen 2004; Kyllönen 2004) or studies focused on parliamentarians (Hylkilä 2003; Berg 2004).

At the beginning of the nuclear debate in 2002, some frustrated Finnish parliamentarians who had taken part in the 1993 debate complained that the nuclear industry seemed to be repeatedly submitting applications until they got the 'right answer' from the Parliament. On the other hand, they complained, the parliamentarians could not prevent the company from reapplying (on the character of the licensing procedure see Chapter 5 in this volume by Säynnässalo). This comment mirrors the frustration of parliamentarians, but it also highlights a fairly common structural feature of decision making about environmentally contested projects, such as new NPP units, where projects often resurface repeatedly. The construction of a fifth nuclear reactor had been discussed in Finland from the 1960s onwards and there had been two previous applications before it was approved – in 1986 and 1992 (Michelsen and Särkikoski 2005). In 1986 the companies withdrew their application following the Chernobyl accident. In 1993 Parliament rejected the decision, but not the process inside the industry to pursue the reactor.

The comment that parliamentarians cannot stop companies from applying is not strictly true, as in theory they could have banned new nuclear reactors by changing the Finnish Nuclear Energy Act. The statement, however, demonstrates the mind-setting of the anti-nuclear side in the debate. The discussion about nuclear phase-down never started after the decision of 1993, nor did environmental organizations seriously try to start one. In effect, the anti-nuclear movement vanished some years after the victory of 1993. The nuclear project, however, never vanished (on the anti- and pro-nuclear lobby see Chapter 8 by Kojo).

The repetition of the parliamentary processes in 1992 and 2002 had several effects on the discussions. First in the decade between these discussions the political landscape changed substantially. Old arguments were demolished or at least diminished in importance: these included the Chernobyl accident, the problem of nuclear waste and the possibilities for alternative energy projects. New topics, such as climate change and the Kyoto emission targets, either emerged or grew in importance. All of these changes were not, however, random or general societal changes. The most important obstacles to the advancement of the nuclear project of 1993, such as the high-level nuclear waste management or the programmes for the development of alternative energy, were actively controlled and steered by the government processes in the period 1999–2001. These processes will be studied below.

Between the DiP processes of 1993 and 2002 several very significant changes took place. In particular, the voluntary workers and the arguments of the anti-nuclear campaign in NGOs and in Parliament seemed to undergo an almost complete change. It seemed that following their victory of 1993 the anti-nuclear movement had, in effect, dispersed by the middle of the 1990s. When a new application for a reactor was submitted in 2000 it led to the formation of a new network of anti-nuclear NGOs (Nissinen 2004). Few, if any, of the veterans of the NGO campaigners rejoined the anti-nuclear movement at that time, even if many of them were still around. This amounted to a major loss of campaign tradition and experience. The new anti-nuclear campaign proved to be quite different from the earlier one and it raised different issues, which possibly has a significant effect on the outcome of the debate.

In contrast with the anti-nuclear movement many of the pro-nuclear actors in the nuclear companies, trade unions and also in Parliament remained the same. However, by 2000 several of the parliamentarians who had been key figures in the pro-nuclear opposition in 1993 had become ministers in the new government. The communication and the outlook of the pro-nuclear campaign were focused on meeting the challenges in the debate. Female parliamentarians, who had been a difficult stakeholder group for the nuclear companies in 1993, were addressed by giving the pro-nuclear lobby several leading female figures: replacing the trade union male lobbyists with pro-nuclear women lobbyists. The threat of climate change was also used as an argument for the development of nuclear power. The pro-nuclear actors could also mobilize considerable resources and support inside governmental organizations, such as the Technical Research Centre of Finland, which provided important research materials, energy books and scenarios for the parliamentarians. In addition, STUK offered almost seamless support for the basic arguments of the nuclear companies. Utilizing the state research organizations proved to be an effective way of lobbying, one with which the NGOs could not compete. The structural aspects of the nuclear project will be discussed later.

The overall landscapes of the debates in 1992–93 and 2000–02

There were substantial differences between the parliamentary debates on nuclear power that were held in 1992–93 and 2000–02. A comparison shows that they were different in tone and topic to such an extent that it is difficult to have a full understanding of the 2002 discussion without seeing the connection and even the polarization in common with the 1993 debate. The discussion was dominated by the anti-nuclear parliamentarians in 1993 and by the pro-nuclear parliamentarians in 2002. In the latter discussion the topics of the discussion moved dramatically away from problems of nuclear energy towards economic arguments about the cost of nuclear energy compared to alternative energy sources. This had implications for the outcome of votes. The major factors, which are argued to be important for the changes between debates of 1993 and 2002, will be introduced below.

The relation of the debates can be seen not only because many of the parliamentarians in 2002 explicitly referred to the debate in 1993 in a frustrated manner. As mentioned above, there were also policy processes, which were features of the 1993 debate and which changed the landscape between the debates. These included, for example, the DiP on the final disposal of spent nuclear fuel and the government climate strategy in 2001, both of them prepared by the Ministry of Trade and Industry and decided in Parliament before the DiP of the new NPP unit in May 2002. One could argue that these policy processes addressed the most important anti-nuclear arguments advanced in the 1993 debate: the nuclear waste problem and the options for alternative energy. These policy papers, debates and decisions provided a new framework for these topics, which proved to be more favourable to the pro-nuclear arguments.

Thus the climate strategy and nuclear waste DiP targeted the most important arguments of the nuclear critics in 1993. In this sense they can be seen as policy-level answers to the arguments of 1993. It is possible that nuclear waste and climate processes were planned deliberately to pave the way for a discussion of the fifth reactor. Many of the pro-nuclear actors in Parliament in 1993 were indeed in governmental positions to start and direct these processes in 2000–01. The pro-nuclear parliamentarians from the opposition Social Democratic Party, who suffered a bitter defeat in 1993, were the responsible ministers in the government in 2000–02, during the preparatory processes of the climate strategy and the nuclear waste DiP. In addition, the energy companies contributed to the effect of preparatory processes, through their careful timing of applications for DiPs and their determined argumentation.

It is very debatable just how much of the dynamics can be credited to these persons and a deliberate pro-nuclear campaign. Over the course of the decade certain new social dynamics emerged in Finnish society. Changes in discussion were taking place in the government and media, which contributed to the creation of a more positive discussion environment for the new NPP unit project. One of the most important aspects is climate argumentation, which will be examined below. Indeed the adoption of the government's climate strategy or some of their decisions about the disposal of nuclear waste would scarcely have been possible in the absence of these wider changes. One of the most remarkable changes was in the media, especially at the main editorial levels. According to polls conducted in 2001, the vast majority of chief editors of the main Finnish newspapers were pro-nuclear. This supports the grassroots experience, reported by NGOs (Nissinen 2004). The participation of researchers in the energy debate also differed between 1993 and 2002. Whereas many researchers had been vocally anti-nuclear in 1993, very few chose to speak out critically of nuclear power in 2002. This is part of the trend of reduced opposition to nuclear power in Finnish society between 1993 and 2002. The possible background reasons for this are listed below.

Nuclear waste DiP

Nuclear waste was the most frequently mentioned problem in relation to nuclear energy in the 1992–93 discussion. During this period the transport of nuclear waste to Russia was banned as the result of an amendment to the Nuclear Energy Act. After this the nuclear companies, TVO and IVO, started to co-operate over the storage of waste in Finland. In 1999 the nuclear waste company Posiva Oy, owned by the nuclear companies, applied for a DiP of a nuclear waste deep repository in Eurajoki. The application was supported by STUK and was approved in government in December 2000. In May 2001 Parliament ratified this decision (see Chapter 6 by Kojo). This was not the final decision to approve the repository, only the first step. But the decision seemed to bury the nuclear waste from the political agenda, drastically reducing the importance of the topic in the 2002 nuclear debate.

In 2001 the nuclear waste DiP was presented by the Minister of Trade and Industry, Sinikka Mönkäre, and Posiva as 'a testing permit'. This framing of the DiP received almost unanimous support in Parliament, with there being very little discussion of the environmental challenges (Raittila and Suominen 2002). A year later, however, this 'testing permit' decision was claimed by the same minister to have solved the problem of nuclear waste, before the excavation of the bedrock had even begun. This change of interpretation seemed to puzzle many parliamentarians and possibly caused them to refrain from using the waste argument. This would be understandable – many of the MPs had little understanding of the technical challenges in the waste repository plans.

It is interesting to notice that in 1993 the nuclear waste debate in Parliament was already concentrated on the political challenges of the process, rather than on the technical or safety challenges. Many had argued simply that as there had still been no decision in relation to what to do with the nuclear waste, it was far too early to make any decision about building a new nuclear reactor. Thus in 2001 the problem of nuclear waste seemed to have been more or less resolved at the political level, simply by taking this political decision in relation to waste, which had been highlighted as the main obstacle to the approval of new nuclear power plants. This did not mean that the majority of MPs would have agreed that there was a safe solution, but rather that they could not find a credible way to raise their doubts.

How did vocal opposition to nuclear waste, the predominant problem of nuclear power in the 1993 debate, change into an approval of the waste DiP? A thorough analysis cannot be presented here, but some issues which relate to the NGO dynamics can be raised. A lack of NGO campaign resources might be one of the factors behind the dampening down of the debate. After a period of relatively high resources and active movements at the beginning of the 1990s, organizations began to experience serious resource problems. During the years 1997–99 there was only one NGO campaigner working with the anti-nuclear waste campaign. Later in 1999 even this campaign was stopped because of financial problems. During this time the parliamentary group of the Green League, who had been opposing nuclear waste plans, ended their objections to the DiP. The Eurajoki municipality accepted the site after a compensation package had been signed with TVO and Posiva (see Chapter 6). The municipality was in considerable debt to TVO, which were then written-off following the acceptance of the repository site. The selection of Eurajoki ended the anti-waste movements in other potential municipalities. The decision by the municipal council of Eurajoki marginalized the local movement still further.

In November 2000, the nuclear waste NGO campaign restarted, with one part-time campaigner.[3] By this time the waste DiP had already been nearly agreed in the government, including the Minister of Environment, Hassi (the Green League). The new campaign had scarce resources, and few links to the previous movements and it was unable to provide the parliamentarians with easily understandable critical material, compared to company and state materials of the nuclear waste plan. The campaign material proved to be too specialized and it was also issued too late to affect the general level of discussion in Parliament. The discussion seemed not to be about the challenges in this particular nuclear waste project, but on the general argument of whether or not nuclear waste was to be buried. The Environment Committee of Parliament invited contributions from 33 organizations – only two of these were NGO campaigners that were critical of the plans. The DiP was approved, with just three parliamentarians voting against the plan.

It is notable that in the Winter of 2000–01, almost half a year before the 9/11 attacks, the Finnish parliamentarians were suddenly very concerned about the fact that nuclear waste rods, lying in their cooling ponds, were vulnerable to terrorist attacks. Less than one year later, after the actual attacks, this fear had vanished suddenly. The topic was not raised when parliamentarians decided to approve new nuclear power, which would expand the time of keeping rods in those ponds from 20 years to perhaps 80 years. Nobody raised the question, not even the NGOs.

The lack of an active nuclear waste campaign by the NGOs had contributed to the situation in Parliament, which then had an impact on the arguments of the debate surrounding the construction of a new NPP unit. The nuclear waste decision and the overwhelming parliamentary support also discouraged NGOs from using the nuclear waste argument in the debate which was to follow.

Climate strategy

Another stumbling block for the reactor project in 1993 was other energy options such as natural gas, renewable energy and energy efficiency, which the government had overlooked at the planning phase. In the 1992–93 discussion Finnish MPs demanded that the potentials should be extensively mapped out. In 1999 the government prepared a promotion programme (MTI 1999) for renewable energy and energy efficiency, which also included the publication of official figures outlining the potential of each energy form in Finland, along with official targets for their development. Many NGOs regarded the targets as modest, although they were at least recognized as some sort of improvement.

After the Kyoto Agreement in 1997 the government found itself in a new situation, involving binding emission targets. In 2000, the government officially started to prepare its climate strategy to meet Kyoto targets. The climate strategy (MTI 2001) framed the debate about energy alternatives into two distinct energy scenarios. The strategy, prepared by the head of the MTI, limited the analysed 'viable' policy options to an increase in nuclear power or an increase in natural gas, with all other emission reduction measures remaining the same. Both scenarios detailed the same, rather limited increase in renewable energy and energy efficiency, which the Minister of Trade and Industry argued were at their maximum potential. This meant that the earlier targets for renewables and efficiency from 1999 were imported into both scenarios without any revision. Although the civil servants in the Ministry of Trade and Industry produced the energy scenarios there was a steering group consisting of different ministers. The Minister of Environment, Satu Hassi (the Green League), who was a member of the group, accepted the scenarios as a compromise.

In 2001, these scenarios were approved in Parliament, although the main committee of Parliament demanded a third scenario based more substantially on renewables. No such scenario was forthcoming from the MTI. Then, one year later, during the debate even the natural gas option was declared by the same Minister of Trade and Industry, Sinikka Mönkäre (Social Democratic Party), to be a 'non-option'. The problems hidden in the natural gas scenario, such as the security of supply and the cost of gas, were highlighted by the minister, who only a year earlier had assured parliamentarians that they had two real options to choose from. On top of this, the energy-saving potentials in the nuclear scenario were doubled overnight in the nuclear option by the MTI, just before the DiP vote on new NPP unit in the government. The reasoning behind this sudden 'extra' potential was that the money saved from the purchase of natural gas could be used to increase investment in energy efficiency.

The Minister of the Environment disagreed with the analysis provided by the MTI, and ordered research to be carried out by the Government Institute of Economic Research (VATT), to compare the different scenarios and saving potentials inside them. This action by the Minister of the Environment was widely criticized both inside government and in the news media. The episode highlights the monopoly of energy scenario production enjoyed by the MTI, which was in many eyes both justified and taken for granted.

The energy scenarios had a significant impact on the discussions. They framed the climate strategy discussion with analysed information and visual materials, graphs and calculations, which were used as background data by the actors in the debate, including by many of the NGOs. Even if the background assumptions of the graphs were criticized, the acceptance of the scenarios as the main topic of discussion tied the participants of the debate to the producers of this information, who were by no means neutral in this matter. There were very few organizations in Finland which had the statistical data and computer models to produce additional energy scenarios. These organizations acted as gatekeepers of what scenarios could be produced, for example by repeatedly ignoring the parliamentary or NGO calls to produce a third scenario based on further increases in renewable energy and energy efficiency. During the nuclear debate, when the natural gas scenario was presented as 'non-option', it was already too late to request the production of new scenarios.

Climate awareness hypothesis

It has been argued, by Berg (2004) and Kyllönen (2004), that an important change behind the nuclear decision of 2002 was the increasing awareness among Finnish parliamentarians of the threat of climate change. The climate threat had, of course, affected the nuclear power debate, dwarfing the perceived problems of nuclear energy. But within this general trend the detailed picture also proves to be both more complex and more surprising.

The 1992–93 parliamentary discussions seem to challenge the 'climate awareness' explanation of the 2002 debate, at least to some extent. In 1992, even before the Rio climate treaty, emission reductions of 60–80 per cent were demanded in the Finnish Parliament by many anti-nuclear MPs. Later in the 2000–02 debate demands for emission reductions were much more modest, meeting the new Kyoto targets. For Finland this meant freezing emissions to 1990 levels by 2010. Compared to 2001–02 the discussions ten years earlier reflected a need for much larger changes in the energy system and living standards in the minds of parliamentarians. In 2002 it was enough to meet the Kyoto target, and this could be done with a reactor plus some other additional measures.

It seems that more important than the simple linear increase of awareness of the climate threat was the reframing of a general and abstract climate challenge to a specific short-term Kyoto emission target, with its own costs and measures. This also allowed the fifth nuclear reactor to be framed as a partial solution to the climate question, which was interpreted by many in the 2002 debate as 'first meeting the Kyoto target'.

However, the awareness of climate change most likely had an effect on the new generation of NGO activists in the late 1990s, who formed the core of the new nuclear campaign actors. For them climate was one of the most pressing environmental issues and they had little in-depth knowledge of nuclear safety or the nuclear waste issue. For this younger generation the looming climate threats seem to have framed the traditional risks of nuclear power in a new fashion. This could have led to the change in argumentation which is analysed below. Instead of focusing on the direct threats of nuclear energy, the NGO message concentrated on climate scenarios, and more sustainable alternatives to emission reduction measures than nuclear power.

The role of environmental NGOs in the parliamentary debate

The role of environmental NGO activities in these debates has been a sub-theme of the analysis. Obviously, many unofficial processes around Parliament played important roles in the opinion forming of the parliamentarians, and these processes were often referred to in the parliamentary debates. Many interest groups announced their support or opposition to the project in attempts to affect the parliamentarians: the labour unions, environmental NGOs, party youth organizations, and constituents. In addition, some research organizations, such as the Technical Research Centre of Finland, were taking an active part in the policy process, in 2002 taking a pro-nuclear stance and effectively providing background materials 'to lobby' parliamentarians. The committee expert hearings acted as filters, determining to which arguments the parliamentarians would be exposed. The decisions about which organizations should be invited to the hearings was made in the committee with some political bargaining. It was also important which persons the invited organizations chose to send, as many of the renewable energy experts and researchers were in effect prohibited from attending the parliamentary hearings and were replaced by their pro-nuclear superiors.

During the two debates the networking of pro-nuclear and anti-nuclear actors has obviously had both many similarities and many differences. On both occasions the industry and trade union representatives strongly supported the construction of the reactor. By contrast, in addition to environmental organizations many of the youth organizations of political parties were in opposition on both occasions. During both debates parliamentarians from the three biggest parties were divided, the National Coalition Party being mostly in favour, with the Social Democratic Party and the Centre Party quite evenly divided. Smaller parties such as the Christian Democratic Party, the Swedish Speaking Party, the Left Alliance and the Green League were mostly in opposition to the new NPP unit.

The most striking contrast of networking between the debates is that in the period 1991–93 the anti-nuclear movement had much broader support across the political spectrum, albeit the official positions of the most important actors and parties remained the same during both processes. Thus in 1993 environmental NGOs in the anti-nuclear campaign were co-operating with the youth organizations of the National Coalition Party, although the National Coalition Party, their mother organization, officially had a very strong pro-nuclear position. In contrast, in 2000 the newly built anti-nuclear movement had lost much of its supporting network actors in the political youth organizations and local trade unions, and even grassroots activism was more marginalized in the debate, despite its relatively large volume.

In 1992–93 there was a group of anti-nuclear MPs actively meeting and coordinating the peer-to-peer lobbying inside Parliament. In 2001 a similar group was formed, but it was not very active, and it did not include many of the most active proponents from the 1993 debate. On the other hand there are reports that in 2000–02 the pro-nuclear side had an active group of MPs, lobbying for the new unit. Hylkilä (2003) has described how this group operated inside Parliament, producing arguments and inviting pro-nuclear lobbyists to meet them in Parliament.

Change in the campaign messages of NGOs

Between the two debates there was a significant and interesting change in the argumentation of the environmental NGOs. In 1993 the NGOs had focused on nuclear safety and the problems inherent in nuclear power, such as nuclear waste. In 2002, they concentrated more on the technical and economical alternatives to nuclear power, such as energy efficiency and renewable energy.

In 2002, the NGOs also made more use of economic arguments than ethical arguments. They argued that nuclear power was not as economic as presented by the pro-nuclear lobby, and that some major costs had been overlooked. All in all the argument was that renewable energy and efficiency are the sustainable answer to emission reductions in the energy sector. One of the main problems with a new NPP unit, as presented by the environmental NGOs, was the claimed negative effect it would have on the development of renewable energy sources. Thus in the final decision of Parliament in Spring 2002 a positive nuclear decision was linked to a new statement requiring more subsidies for renewable energy and legislation to establish compulsory energy saving.[4] As the inherent safety or moral problems of nuclear power itself were not highlighted in the discussion, many parliamentarians evidently concluded that supporting renewable energy would take care of the most pressing problems of nuclear energy.

It is credible that the change in the type of arguments advanced by NGOs may have affected the discussion in Parliament, because the NGOs were one of the most active groups in the production of anti-nuclear background information for parliamentarians. Those parliamentarians who were critical of nuclear power often used this material to extract arguments for their speeches. In fact in 2002, few environmental NGOs produced material about nuclear safety or nuclear waste, which meant that those MPs who were critical of nuclear power had less information on which to base their arguments. Moreover, the economically and technologically focused communication of NGOs placed them in an awkward position, when their main points were downplayed by professional energy economists and even pro-renewable organizations. It can be argued that environmental NGOs could not compete in social credibility with these organizations outside their traditional field of expertise. In contrast, in 1993 NGOs concentrated more on the moral problems, safety and environmental risks of nuclear power, where their credibility was higher.

How, then, did this change of communication come about? Was it a deliberate change of tactics by NGO actors, or entrainment, an uncontrolled social dynamics?

Processes behind the formation of NGO messages

It is worth considering to what extent the change of NGO messaging was affected by the changes in the overall discussion in Finnish society. For example, it has been pointed out that in 2002 the memory of the Chernobyl accident (1986) was already much less vivid than it had been in 1993. It has also been argued that an increased awareness of climate change altered the priorities of many participants in the debate, possibly including the NGO actors. Certain policy processes, such as the nuclear waste DiP, could also have hindered the critics from using, for example, their favorite nuclear waste argument. The author states that these changes all had a role in the changes of the NGO communication, and produce a dynamic which produced the change in the communication.

The change in the NGO communication between the debates of 1993 and 2002 was not a gradual one; it coincided with the drastic changes in the persons and organizations in the anti-nuclear movement, which happened during the period 1995–2000. There was an almost complete discontinuity of organizational transmission of campaign materials and narratives in the NGO nuclear campaigns between 1995 and 1999, which can be argued to have had significant implications for the campaign and later the parliamentary debate. This change of actors coincided with the political processes, so that for example in 1999–2000 during the preparation of the nuclear waste DiP there was no active campaign against the nuclear waste DiP, because of the lack of NGO financing and actors. This lack of an anti-waste movement thus also affected the nuclear waste DiP discussion.

Repetition of the process and the civil society

One of the most interesting things in the repetition of the DiP process was the effect it seemed to have on the voluntary anti-nuclear campaign in NGOs and in Parliament. Few, if any, of the veterans of the NGO campaigners rejoined the anti-nuclear movement at that time, even if many of them were still around.

The change of actors was also very striking on the anti-nuclear side in Parliament. Many key persons who had been active in the 1993 debate were not very active at a later date. Few of them had actually changed their opinions, but it seemed that they did not want to take up the job again, or to be associated strongly with anti-nuclear campaigning. It is difficult to estimate whether this phenomena was linked to fatigue brought by the repetition of the tiresome campaign process, or whether it was perhaps a sign of the increasingly positive attitudes towards nuclear power across society in general. The fact that many of the veteran MPs or NGO campaigners still encouraged others to act against nuclear power, even if they themselves were not very active, suggests that fatigue was a considerable factor at this time. Among these figures was the key figure of the anti-nuclear campaign in 1993, MP Matti Vanhanen of the Centre Party. In 1992 Vanhanen had handed in the amendment which rejected nuclear in the official energy policy. In 2002 he was not active in most important anti-nuclear MP groups.

This loss of actors on the anti-nuclear side is in stark contrast with the employed industry lobbyists, trade union campaigners and civil servants of MTI, many of whom had remained in the same organizations throughout this period. Fatigue did not seem to have the same impact on the pro-nuclear side of Parliament. Many of the most important pro-nuclear proponents of the 1990s again played a crucial role in 2002. It even seemed that the bitter defeat of 1993 had energized some key people into action seven years later.

Discontinuity of campaign tradition in the NGO campaign, 1995–99

The pro-nuclear campaigners in the industry and labour unions had more experienced actors, a wider transmission of accumulated knowledge and experience, and a better network both inside Parliament and in society. In contrast, there was an almost complete discontinuity of organizational transmission of campaign narratives in the NGO nuclear campaigns between 1995 and 1999, which had significant implications for the debate.

In effect, the previous anti-nuclear NGO network had already ceased to exist by 1995. The transmission of the campaign tradition had almost been stopped in 1997, when one of the most active anti-nuclear NGOs, Greenpeace Finland, temporarily closed its office, as a result of a financial crisis. Many old campaigners did not return to work in the newly formed Greenpeace Nordic and there was relatively little contact between the old and the new campaigners. The Greenpeace office in Finland only reopened in 1999. Only one experienced Greenpeace nuclear campaigner continued to work for the nuclear waste campaign of another NGO, the Finnish Association for Nature Conservation, a campaign that ended in 1999. In addition, in 1997 and 1998 there were two tours organized by Friends of the Earth Finland to raise the issue of opposition to uranium mining. This, however, did not bring many new actors to the nuclear campaigning, nor did it produce a long-term campaign, which would have continued until 2000.

By the end of 1999 only one person working in the Finnish NGOs was working on the energy campaign – a part-time energy campaigner in the new Greenpeace Helsinki office.

The new nuclear campaign had to start from almost nothing. The campaign strategy, argumentation and the materials were 'reinvented' during this time, and they also went through a profound change compared to the style of the 1993 campaign. There was a strategic change in the arguments presented against nuclear power: the main message moved away from an emphasis on the risks of nuclear power to stressing the benefits of renewable energy, which was threatened by the nuclear project. The campaign style was changed, and a greater weight was placed on reports and technical argumentation than had been the case in 1993. Greenpeace was one of the main actors starting up the anti-nuclear network in 2000. The campaign messages of other NGOs, with their new generation of actors and their newly formed anti-nuclear network, seemed to follow the lines of the new Greenpeace campaign, forming new lines of argument for the whole movement.

Even if there was a lack of continuation in the nuclear movement, the traditions of this new kind of campaigning can be traced back to earlier NGO movements in Finland. Many of the energy campaigners of this time, including the Greenpeace energy campaigners in the period 1999–2002, came from an environmental NGO called 'Dodo – the Living Nature for the Future'. Dodo had been founded in 1995 as 'an alternative NGO for ordinary people', which aimed to avoid the traditional confrontations of the environmental debate. It had a distinctive style of matter-of-fact argumentation, an emphasis on technological solutions, an emphasis on climate and an explicit dislike for traditional demonstrations or campaigning.

Dodo, which received most of its support from university students, can be seen as challenging the traditional image of the environmental activist. The narratives of Dodo emphasized the importance of discussion, as the way to come up with novel solutions to solve old environmental conflicts. In 1998 Dodo members were also the founders of the first customer-owned wind power company in Finland. Dodo-style campaigning highlighting positive environmental solutions could have affected the change of campaigning style in the nuclear campaign in 1999. The importance of this campaign tradition is highlighted by the fact that in 2002 four out of the five main anti-nuclear NGO campaigners had previously been on the board of Dodo and had no previous NGO experience. In effect, most of the main campaigners in new energy campaigning came from a single organization.

However, there is one notable exception to this situation: a continuum of campaigners from 1993 to 2002 (Nissinen 2004, 88–91). They were between three and ten elderly campaigners, whose campaign had continued from 1986, and had concentrated on the risk and moral aspects of nuclear power. In fact, during some phases between 1997 and 2002, they were the only group to focus on the risks of nuclear power in campaign messages. It has been speculated by some NGO actors that their very different socioeconomic profile and their grassroots style of communication marginalized this type of traditional and risk-focused campaigning, in the eyes of new NGO actors. This might then have further discouraged the other NGOs from adopting the risk argumentation.

The relative absence of moral arguments in the 2002 debate is interestingly mirrored against this dynamics of NGOs. After all, moral arguments need few resources to be produced. The use or non-use of moral arguments against nuclear power in the period 2000–02 was probably affected by many factors. One of them was that people using moral arguments were dragged repeatedly into a technological debate. Thus, if NGOs argued about the moral problems of producing nuclear waste or exposing public to the risk of nuclear accidents, they would be answered by technical arguments that explained why these risks do not exist. This lured some NGOs to enter arguments about technical problems.

The pro-nuclear lobby emphasized this effect by actively branding anti-nuclear NGOs, claiming that they were using 'scare tactics' and 'emotional campaigning', which highlighted the dangers of exploding reactors and leaking nuclear waste, based on no real knowledge of the risks. Those NGOs that made the greatest use of the moral and safety arguments did not succeed too well in the technical debates that followed. Even if successful, the safety debate was often too technical for MPs to follow. The trials to use the moral arguments by major NGOs, for example showing the health effects of documented nuclear accidents in parliamentary hearings, were strongly denounced as scaremongering. These incidents meant that moral arguments could not be easily used effectively without engaging in the technological debate.

Conclusions

In the 2000–02 the NGO campaign messages were focused more on the technical possibilities and economics of renewable energy sources than on the risks of nuclear energy, even if some reference was made to the risks. Although some actors discussed this change of focus, at the level of the movement as a whole it does not seem to be a deliberate decision, but rather than an effect of entrainment of argumentation. The situation affected the debate, as almost no NGO highlighted the risks as a primary reason to oppose the new NPP unit. NGOs produced relatively little 'alternative information' about reactor safety or nuclear waste, and concentrated instead on the issue of renewable energy. This effort leads NGOs into unfamiliar fields of discourse, such as economics, technology and jobs, where the companies, labour unions, and researchers, had considerably more credibility.

The long timeline and repetition of DiP debates in 1993 and 2002 highlights the non-symmetry of the two sides of the nuclear debate. On the one side there were the professionals of the pro-nuclear establishment in the companies and state organizations, and, on the other, financially insecure environmental organizations, which were at times forced to stop campaigning. This setting can have larger ramifications for the Finnish environmental debate, decision making and its background assumptions.

Environmental policy makers in Finland often seem to take for granted that they have different sides of the environmental debate to listen to. The dynamics of the 1993 and 2002 anti-nuclear movement proves this position problematic, even following the victory in 1993. Even the most important asset of the Finnish environmental organizations – their in-depth knowledge of the nuclear issues – was in effect lost.

The basic challenges were stated at the beginning of the chapter: which organizations are providing the competing information to frame and fuel the environmental debates, and do they have the resources to perform this role effectively? In the Finnish nuclear debate of 2000–02 the task of providing alternative information or arguments for the policy makers was, to a large extent, left to the NGOs. The few efforts by the Minister of Environment to utilize state research centres for this purpose were blocked and led to outcries from the government and media. By contrast, NGOs had very limited financial resources to provide the information.

It is an interesting detail that before the nuclear debate in 2001 some NGOs had been granted funding by the Ministry of Trade and Industry to take part in the energy debate (Kojo 2004b, 249). The small size of these grants, totalling approximately €43,000, however, highlights the asymmetry of resources even further. One reason for the financing is that the debate as such was seen as being important for the formal decision-making process. The process was later touted internationally as providing a model for nuclear decision making.

The cases at hand show that, rather than looking at the formal hearing processes, the focus of any examination should be on the backstage of the debate, starting from a consideration of who is producing EIAs and who is commenting on them. As noted earlier, the role of research organizations – and especially STUK – is interesting in both nuclear waste and nuclear power debates when it comes to the provision of critical information.[5] Generally, STUK had been supporting both DiPs and did not highlight the challenges in the projects in their presentations to Parliament. The research materials of STUK did pose critical questions, but always in a form which ended in a positive and optimistic projection that they were capable of being solved. The organization has no resources to carry out its own confirming research and its communication is very close to Posiva and the two nuclear companies. This dilemma is not unique to Finland, but needs further analysis especially now, when new NPPs are being built. An analysis of how the state research organizations see their own role in environmental debate could be particularly illuminating.

The challenging question of democratic pragmatism remains: If environmental civic movements cannot provide the resources for effective long-term high-quality campaigning, how could they compete in a democratic debate process against commercial organizations, which are backed by vastly greater resources? Or when they have to argue against state research institutions, which have the highest credibility in a society like Finland and have close links and connections to the commercial world? The answer is not straightforward, but is rather immersed in different cases. The cases can provide us with examples and counterexamples, and even overall trends, about how the present practices of environmental debate in western societies are working. And some, such as the nuclear debates in Finland, highlight the need to examine the background of the debates and the institutional solutions for providing background information for environmental decision making.

Notes

1. In 1993 Parliament rejected the Decision-in-Principle (DiP) of a new NPP unit. In 2002 the DiP was ratified by Parliament.
2. The aim of the PhD is to study the storylines of the parliamentarians in the nuclear debates in the period 1992–2002. The author took a personal role in the 2000–02 debate as a NGO energy campaigner in Greenpeace Nordic. He has experienced both the decision-making process inside Parliament and the forming of the new anti-nuclear NGO movement in Finland in the beginning of the millennium. The author has closely followed the Finnish energy debate since 1996, as a volunteer NGO campaigner and as an energy policy researcher at the university. This has all, of course, given him a somewhat unique viewpoint in respect of the NGO energy campaigning. It also provided him with a perspective and tone into transcribed debates, and insights into how they relate to the larger dynamics of the campaigning inside Parliament. It is obvious that although the author's interest has been to objectively reflect the dynamics of the debate, his background prevents him from being any kind of a theoretical outside observer. From the methodological point of view the author's position could be described as one of a participative observer.
3. The author was the campaigner.
4. The extra subsidies never materialized, and the energy efficiency law was later abandoned by an official working committee, consisting mostly of energy companies. (See Chapter 2 by Lampinen.)
5. Many of the relevant organizations were under the jurisdiction of the MTI, for example, STUK nuclear division and the Technical Research Centre of Finland.

References

Berg, A. (2004) *Ydinvoiman uudet vaatteet. Kansanedustajan näkökulma vuosien 2000–2002 ydinvoimakeskusteluun* (The New Clothes of Nuclear Power. Parliamentarian's Viewpoint on the Nuclear Discussion of 2000–2002). Unpublished Masters thesis of Journalism and Mass Communication. University of Tampere, Tampere, http://tutkielmat.uta.fi/pdf/gradu00424.pdf.

Dryzek, J. (1997) *The Politics of the Earth: Environmental Discourses*. New York: Oxford University Press.

Hokkanen, P. (2004) *Kansalaiskeskustelu lisäydinvoimasta – mielipidekirjoitukset osallistumiskanavana* (Public Discussion on Additional Nuclear Power Capacity – Written Public Opinions as a Channel of Participation). MTI Financed Studies 6/2004. Helsinki: Ministry of Trade and Industry.

Hylkilä, I. (2003) *Suomen viidennen ydinvoimalan lobbaus kansanedustajan näkökulmasta* (Lobbying the Fifth Nuclear Power Plant Unit in Finland from the Point of View of a Member of Parliament). Unpublished Masters thesis study in Organizational Communication and PR. Department of Communication. Faculty in Humanities. University of Jyväskylä. http://urn.fi/URN:NBN:fi:jyu-2003920889.

Kojo, M. (2004a) Eurajoki – ei paras mahdollinen, mutta poliittisesti sopiva. (Eurajoki – Not the Best Possible, but Politically Suitable). In M. Kojo (ed.), *Ydinvoima, valta ja vastarinta* (Nuclear power, power and resistance). Helsinki: Like, pp. 127–58.

Kojo, M. (2004b) Yhteenveto (Summary). In M. Kojo (ed.), *Ydinvoima, valta ja vastarinta* (Nuclear Power, Power and Resistance). Helsinki: Like, pp. 231–58.

Kyllönen, S. (2004) Ydinvoiman ilmastonmuutos (Climate Changes and Nuclear Power). In M. Kojo (ed.), *Ydinvoima, valta ja vastarinta* (Nuclear Power, Power and Resistance). Helsinki: Like, pp. 51–85.

MTI (1999) *Uusiutuvien energialähteiden edistämisohjelma*. Kauppa- ja teollisuusministeriön julkaisuja 4/1999. Helsinki: Edita.

MTI (2001) *National Climate Strategy – Government Report to Parliament*. Publication 5/2001, Helsinki: Ministry of Trade and Industry.

Lammi, H. (2004) Tarinat kovasta ytimestä (The Stories Concerning the Hard Nucleus). In M. Kojo (ed.), *Ydinvoima, valta ja vastarinta* (Nuclear Power, Power and Resistance). Helsinki: Like, pp. 11–50.

Litmanen, T. (2008) The Changing Role and Contribution on Social Science to Nuclear Waste Management in Finland. *Energy & Environment* 19(3–4), 427–53.

Michelsen, K.-E. and T. Särkikoski (2005) *Suomalainen ydinvoimalaitos* (Finnish Nuclear Power Plant). Helsinki: Edita.

Nissinen, J. (2004) Valtavirran vastarinta (The Resistance of Valtavirta). In M. Kojo (ed.), *Ydinvoima, valta ja vastarinta* (Nuclear Power, Power and Resistance). Helsinki: Like, pp. 87–105.

Raittila, P. and P. Suominen (2002) Keskustelu ydinjätteen loppusijoitusta koskevasta periaatepäätöksestä eduskunnassa ja mediassa (Discussion on the Decision-in-Principle on Nuclear Waste Disposal in Parliament and the Media). In P. Raittila, P. Hokkanen, M. Kojo and T. Litmanen (eds), *Ydinjäteihme suomalaisittain* (The Finnish Nuclear Waste Miracle). Tampere: Tampere University Press, pp. 92–113.

Sjölander, A. (2004) *Kärnproblem. Opinionsbildinig i kärnavfallsdiskursen i Malå*. Medier och kommunikation 7. Umeå Universitet.

Tarjanne, R. and S. Rissanen (2000) Nuclear Power: Least-Cost Option for Baseload Electricity in Finland. Paper presented in *The Uranium Institute 25th Annual Symposium*. 30 August–1 September 2000. London.

Ylönen, M. (1999) The Formation of a Power Relationship in an Environmental Conflict. In E. Konttinen, T. Litmanen, M. Nieminen and M. Ylönen (eds), *All Shades of Green: The Environmentalization of Finnish Society*. SoPhi, University of Jyväskylä, 150–79.

Part III
Decision Making in Finland

4

The Discursive Dimensions of a Decent Deal: How Nuclear Energy Evolved from Environmental Enemy to Climate Remedy in the Parliament of Finland

Annukka Berg

> [The discussion on nuclear energy] was no longer about yes or no, but it gained, particularly through the versatility of this environmental dimension, quite relevant added value.
>
> Centre Party parliamentarian who voted for the fifth nuclear power plant unit

> [I]n a way [the nuclear decision] was also about testing who has power. So in this case the power from the grassroots level lost and the power from the Eteläranta ten [the headquarters of the Confederation of Finnish Industries] won.
>
> Green parliamentarian who voted against the fifth nuclear power plant unit

Introduction

This chapter aims to shed more light on the ways in which the members of the Finnish Parliament viewed the decision on the fifth NPP unit. Why did some of the MPs vote for the building of the new NPP unit? Why did some of them vote against it? In which ways have the general ideas about desirable development, economic growth and environmental risks affected their decisions? What was the structure of the *discursive* field in general? When analysing the discourses the author has also been interested in the ways in which the parliamentarians viewed the role and performance of different *actors* in deciding on the nuclear issue. How did they see their own role? What were their ideas about expert knowledge and citizen participation in the process?

This chapter is based on the author's master's thesis (Berg 2004) and summarizes some of its core findings. The material of the analysis has been gathered by conducting 12 focused interviews with members of the Finnish Parliament. In the selection of interviewees, the author sought to find parliamentarians who had been uncertain about their decision for a long time or who had somehow voted differently from their reference groups or their previous voting records on nuclear issues. The hypothesis suggested that for these parliamentarians, the decision on the fifth NPP unit was harder to make. Therefore, it would be especially interesting to see what strategies they had used to find their way.

What the material shows is a wide variety of logics and arguments that served to justify both pro- and anti-nuclear decisions. In addition, it also reflects the general importance attached to the decision. The building of the fifth NPP unit was, indeed, seen as one of the most important decisions of the entire 1999–2003 parliamentary period. The analysis would suggest that one of the reasons for this could be the deep structure of the discursive field; at the core of the decision stood three distinctive ways of thinking about the desirable development of Finnish society: the simple, ecological and reflexive modernization. These discourses differ from each other, particularly in their viewpoint in relation to environmental risks and the role that they believed should be played by technology and expert knowledge in solving these problems.

The following section discusses the theoretical roots of the chapter and sheds light on the environmental risks of nuclear energy and climate change as well as the roles citizens and experts play in defining and managing these risks. In addition, the three different types of modernization will be introduced in depth, after which the material and the method utilized in this chapter are briefly described. The analysis section discusses the special character of the decision on the fifth NPP unit, the discursive field and the roles of different actors. Finally, the results will be summed up and some further conclusions will be drawn.

Environmental risks, actors and ideal types of development

The risk profile of nuclear power and climate change

Risk is an attribute of decision and an important concept in this analysis as it is something that is taken into account in the course of the decision-making process. Risk refers to the consequences of decisions and also the uncertainty attached to them. The term is used in a number of different ways. In everyday life as well as in the social sciences the use of the term is rather imprecise; by contrast, in the natural sciences and technology its use is rather more restricted. What is common to all of the definitions, however, is the concept of a potentially damaging, harmful or dangerous event, since damage and possibility are the two key components of risk. It is also worthwhile to draw a distinction between ordinary and catastrophic risks. When a risk is catastrophic, the role of probabilities diminishes because it may be the case that even a very improbable risk may be too large to take. Thus, in such circumstances ethical judgements may serve as important grounds for the decision (Kamppinen et al. 1995, 11–15; Malaska et al. 1989, 94–5).

Beck (1992, 21) defines risk as a 'systematic way of dealing with hazards and insecurities induced and introduced by modernisation itself'. As Tapio Litmanen points out (2001; see also his chapter in this book), the process of risk definition and recognition in relation to nuclear power is a strongly social and cultural process. This global nuclear culture consists of knowledge, beliefs, and values that are compressed into interpretative packages for and against nuclear technology. The public perception of nuclear risks and the ongoing debates over their acceptability have gone hand in hand with the birth and development of anti-nuclear protests. On the other hand, in the case of climate change, research – and particularly the work of the Intergovernmental Panel on Climate Change – has played a central role. Thus, in the climate context, social movements have been less important (Tirkkonen 2000, 204).

What affects the reception of environmental problems, risks and their urgency is the existence and experience of other social problems (Suhonen 1994, 122). According to Kojo (2004, 239), one explanation of why the Parliament of Finland approved the DiP to build the fifth NPP unit was that the risk perception of a serious nuclear accident had changed. In addition, a political solution had been found to the issue of the storage of nuclear waste (for more details about the DiP and the site selection process for nuclear waste, see Chapters 3 and 6 by Lammi and Kojo in this book). The Chernobyl accident (1986) had been defined as a special case that could not happen in the western world. Even though a wide variety of other risks were also associated with the decision, from energy dependency to small particle emissions and Russian nuclear safety (Lammi 2004, 42–3), writers such as Kyllönen (2004) find that changes in the environmental image of nuclear power were most decisive in the Finnish pro-nuclear decision of 2002. According to Kyllönen (2004, 52), the significant factors were, first, that climate change had become the environmental problem dominating the public discussion and, secondly, that nuclear power was successfully marketed as an 'emissions-free' energy solution.

The risk profiles of climate change and nuclear technology have both similarities and differences. In comparing the two, the dangers of nuclear power can seem a little clearer, more familiar and geographically more restricted. For example, we already have some experience of one of nuclear power's worst risk scenarios in Chernobyl. Moreover, while Chernobyl was a severe accident, its impact area was still rather restricted compared to that of climate change. What is common to the risk perception of both nuclear energy and climate change, on the other hand, is that neither nuclear radiation nor the gradual change in climate can be sufficiently perceived by human senses alone. Such risks can also be called new as they are no longer tied to their place of origin: the industrial plant. They are what Beck (1992, 22) calls 'the new kinds of industrialized, decision-produced incalculabilities and threats'. What is significant is that averting and managing catastrophes can include a reorganization of power and authority (Beck 1992, 22–4). This is one of the reasons why it is so interesting to analyse the way the members of the Parliament of Finland perceived different actors.

As noted above, one of the ways to deal with risks is to conduct a comparison of them. When questioned in the mid-1990s, the Finnish environmental opinion leaders found climate change to be a more significant global environmental threat than nuclear power (Järvelä and Wilenius 1996). What the leaders saw as particularly problematic in relation to nuclear energy, however, was that it is very difficult to calculate the probability of the risks – an issue that has been on the political agenda for several decades. In the 1970s, the safety of nuclear power plants was recognised as being a worthwhile environmental risk, and during the course of the 1980s the issue of nuclear waste appeared on the agenda. By comparison, the emission problems of fossil fuels only began to be recognized in the 1980s. Back then, however, the discussion focused largely on the sulphur and nitrogen emissions that caused acidification (Kyllönen 2004, 53–4; Vahtera 1986). Meanwhile, the hegemonic climate discourse has evolved since 1990 with the United Nations Framework Convention on Climate Change (1992) and the Kyoto Protocol (1997) marking two of the most important milestones (Tirkkonen 2000, 95). Still, as early as the 1970s, and particularly since the end of the 1980s, environmental reasons were beginning to figure in decisions to opt for nuclear power rather than fossil fuels (Heinonen et al. 1978, 51–2, 161–2, 169; Vehmas 1995, 170).

Risk society and the three ideal types of modernization

In the study undertaken by Järvelä and Wilenius (1996, 57), Finnish environmental opinion leaders noted that the root cause of environmental risks is the way western societies function. This root cause reflects very 'structural' understanding of the reasons behind environmental risks. The same structural understanding is also central to the thinking of Ulrich Beck, whose well-known theory of *risk society* has been considered an important portrayal of our society. According to Beck (1992, 22–3), new risks cannot be restricted socially or with regard to time and place. Technology can be utilized only for minimizing risks, not for abolishing them, because '[a]long with the growing capacity of technical options – grows the incalculability of their consequences'. The rules of attribution and bearing responsibility cease to function, and the calculation of risks, as science and legal institutions have established thus far, collapses. Dealing with the consequences of modern productive and destructive forces in the normal terms of risk is a false, but nevertheless very effective, way of legitimizing them. For Beck, the use of nuclear power plants is a central example of the way risk society works. For example, according to him, nuclear accidents should no longer be considered accidents as they affect future generations that have not yet been born. Nuclear energy is considered a dangerous game with the imputed infallibility of technological development. It requires social controls, the 'authoritarian nuclear state' (Beck 1992, 178).

For Beck (1992, 231–2) 'reflexivity' describes a different epoch of the modernization process. The term refers to development in which monopolies emerging with industrial society and built into its institutions break up. This is happening in the case of science, for example, which has enjoyed a monopoly over rationality. Beck (1992, 234) highlights the importance of self-criticism: 'Things that only now have been able to make their way with great difficulty against the dominance of professions or operational management must be institutionally protected.' What these institutionally protected forms of self-criticism could mean are alternative evaluations, alternative professional practice and discussions within organizations and professions of the consequences of their own developments. For Beck (1994, 2, 5–6), simple modernization means the disembedding and re-embedding of traditional social forms by industrial social forms. Thus, *reflexive modernization* can be seen as the disembedding and re-embedding of industrial social forms by another modernity. It entails self-confrontation with the effects of risk society that cannot be handled within the system of industrial society. Reflexive modernization is a development idea that criticizes industrial society and emphasizes the capability to learn. In its processes, sub-politics, or the shaping of society from below, play an important role (Beck 1994, 23). Moreover, in the core of reflexive modernization lies a theory of the ever-increasing powers of social actors. This theory includes a programme of individualization as well as one of empowerment (Lash 1994, 111–13).

A concern with issues of ecology is a common focus for writers of reflexive modernization (Beck et al. 1994). Therefore, ideas about *ecological modernization* provide an interesting alternative. Ecological modernization holds that the environmental challenge cannot be faced with simple end-of-the-pipe solutions. Rather, it suggests changes that are less institutional, radical and thorough than those of reflexive modernization. Tirkkonen (2000, 29) defines ecological modernization as a process of change in which modern society develops through preventive, innovative and structural acts towards a more ecologically acceptable society. Ecological modernization answers the environmental challenge by integrating it into the modern project already launched by the thoughts of the Enlightenment. As Spaargaren (1997, 12) summarizes it, ecological modernization is about highlighting the possible win–win solutions between ecology and economy. Moreover, it redefines the relationship between the state, its citizens and private corporations. Ecological modernization points out that, as many of the most challenging environmental problems of our time extend beyond the borders of nation-states, there exists a need for a supra- or transnational approach. Thus, ecological modernization recognises the structural character of environmental problems, but still differs essentially from radical green perspectives. For example, Hajer (1996, 253–5) notes that ecological modernization has also been criticized for being a technocratic project. From this point of view, ecological modernization is more appropriately seen as a repressive answer to radical environmental discourse than its product.

This chapter views ecological modernization as a kind of compromise field between simple and reflexive modernization. It shares with reflexive modernization the idea that the environmental challenge requires serious action. On the other hand, its roots in the Enlightenment, and its optimistic attitude towards science and technology, resemble those of *simple modernization*. Both simple and ecological modernization concentrate power over knowledge into the hands of scientific experts. What is characteristic of simple modernization is that it plays down the environmental and other risks of industrial society. It sees risks as technical questions and counts on modern technology and established experts to solve them. As Lash and Wynne (1992, 2) put it, 'Theories of "simple modernisation", from Habermas to Marx to mainstream Parsonian sociology, all share a sort of utopic evolutionism, whether its motor be communicative rationality, the development of the means of production, or structural differentiation and functional integration.'

According to Lammi (2004), the future of Finnish society was an important theme in the 2000–02 nuclear discussion. In addition, Sjölander (2004) found that the discussion about nuclear waste management in Sweden dealt with a complex web of essential issues for society today, something she calls core issues. These results explain why it is worthwhile to conduct an exploration of the development ideas and 'core issues' behind the nuclear discourses of the Finnish parliamentarians. Moreover, the hypothesis suggested that while environmental arguments had become increasingly legitimate in nuclear discussions, the power that the environmental movement still held after the Chernobyl accident and at the beginning of the 1990s may have become lost in the nuclear debates of the twenty-first century (see Chapter 3 in this book for more details). On the basis of the theoretical literature presented above, one possible explanation for this could be the growing importance of ecological modernization over that of thoughts on reflexive modernization. This theme will be further explored in the analysis below.

Citizens and experts in the fields of energy policy and environmental risks

The ideals of development presented above differ in the way they relate to expert knowledge, technology and the role of citizen participation in societal development. In this subsection, this understanding is deepened further by examining the roles given to experts, counter-experts and lay people in discussions on environmental risks, energy policy and nuclear power.

When analysing the actors, *power* is a central concept. According to Foucault (2000, 338–40), the exercise of power is a way in which some act on others. Power relations are exercised, to an important extent, through the production and exchange of signs. Thus, power functions through discourses and is realized, for example, when experts define environmental risks related to energy policy. Science converges with politics, and conflicts and arguments easily develop from political to technical. Thus, in questions and cases in which the role of expert knowledge is potentially strong, it is important to examine definitions and discursive practices. Because science plays an important role in defining the environmental risks of nuclear power and climate change, the decision on the fifth NPP unit represents just such a case.

According to Kamppinen et al. (1995, 174), an *expert* is someone who knows about a thing more than do members of the general population. When the division of power deepens, there is a growth in the variety of experts with different sectoral and interest backgrounds. Control over the risks of energy policy has traditionally remained in the hands of experts with specialized education. In Finland, however, one problematic aspect is that expert discussion is split into levels and fields that are poorly interconnected. This practice threatens to prevent the birth of heterogenic expert circles that the theories of ecological and reflexive modernization would consider to be important for the development of society (Kantola 2004, 110). In general, as is pointed out by Litmanen in this book, Finns value things such as Enlightenment, state, bureaucracy and technology. Even though mutual trust among Finns has somewhat diminished in the twenty-first century, it nevertheless remains quite strong (Ilmonen and Jokinen 2002, 213).

Lay people can be seen to constitute a counterweight to experts as they offer no specialist knowledge of the issue at hand, at least not in institutionally defined terms. On the other hand, in modern democracies civic movements and other stakeholders are considered to be important players in policy-making processes (Kamppinen et al. 1995, 173). In Finland, however, the situation is that the decision making on energy issues is considered distant, and the opportunities to influence it slight. As many as two-thirds of Finns feel that citizens' opinions have been insufficiently heard in energy decisions (Energy attitudes 2006, 2007). Even though the question of nuclear power has aroused widespread civic activity in Finland and abroad, many structural features of the field make citizen participation difficult. Even observing risks related to climate change and the use of nuclear energy requires scientific knowledge. Nevertheless, the production and use of expert knowledge is not free of conflicts of interest. For example, Kyllönen (2004, 52–3) emphasizes the role of the governmental preparation process in the Finnish national climate strategy as an important way to bring the nuclear alternative into the new environmental policy context. If citizens wish to participate in the discussion, they must often adopt the language and presumptions developed by others. According to Hylkilä (2003, 158), in the 2000–02 nuclear discussion, the anti-nuclear actors had to act, at least to a certain extent, on the terms of the pro-nuclear camp.

Kojo (2004, 234–5) reminds us that the question of nuclear power has been on the Finnish political agenda for around fifty years. Critical discussion, however, began in the country a bit later, in the 1970s. The activity rate of the *anti-nuclear camp* has been changing over time based on the number of nuclear power building projects, the number and severity of accidents and trends in the international movement. When the plans to build the fifth NPP unit were announced in 2000, it signalled the rise of a new anti-nuclear movement in Finland. A network called Valtavirta (Engl. Mainstream) emerged with the goal of renewing nuclear-related citizen activities and reviving the movement's credibility. As Lammi points out in this book, during the 2002 decision on the fifth NPP unit, Finnish NGO messages laid a particular emphasis on the technical and economical alternatives to nuclear power. The goal was to present the 'best arguments' as regards the environmental and economical effects of different energy models (Nissinen 2004, 100–2). This strategy contrasted with the 1993 nuclear combat in which NGO messages were more closely related to nuclear safety and to other inherent problems of nuclear power such as the question of nuclear waste. What was special about the 2002 debate was that pro-nuclear citizens also organized themselves, both locally and nationally. This led to the emergence of the pro-nuclear youth movement 'Ydinenergianuoret' and the local 'Pro Loviisa' movement in Loviisa, a community with two nuclear power plants. The fact that there were now citizen movements in both the pro- and anti-nuclear camps affected the way the case looked from the outside. As a result, public debates were no longer polarized between the pro-nuclear industry and the anti-nuclear citizens; citizens' groups participated in both camps (Kojo 2004, 237; Rosenberg 2004, 170).

According to Hylkilä (2003, 116, 136–9), in the 2002 discussion on the fifth NPP unit, citizen feedback influenced the decision of only some of the MPs. Meanwhile, professors, ministry representatives and other institutionally strong actors were generally considered to be the most influential of the information sources. The way the issue was handled in the Parliament and in the parliamentary groups had a substantial effect on the opinions of the parliamentarians. In addition, the actions of interest groups such as lobby organizations for industry, environmental NGOs and labour unions also had some influence on events. However, the information they offered was in general considered to be less trustworthy than the information delivered by so-called independent actors such as the scientific and government representatives. As institutionally strong experts were well received, they served both pro- and anti-nuclear camps. The fact that people from universities, industry and energy companies were involved in the anti-nuclear coalition gave more credibility to the camp's image. For many MPs, however, the profile of the anti-nuclear camp still remained too closely aligned with the Greens and the radical movements, and some considered the camp's identity to be rather unfocused.

Materials and the method

Interviews with 12 parliamentarians

The empirical material for this study has been gathered by using *focused interviews*. The focused interview is a half-structured interview method first developed by Merton et al. (1956). In the research process, six themes formed the outline of the interviews. These themes dealt with the positions of the parliamentarians vis-à-vis nuclear energy; the risks and possibilities associated with the use of nuclear power; the values, views and expert knowledge that factored into the decision on the fifth NPP unit as well as the ways the parliamentarians perceived the role of the media and the debate in general.

As mentioned above, the research material comprised 12 interviews with members of the Finnish Parliament. There are a total of 200 parliamentarians in the Parliament of Finland, so the number of MPs interviewed represented 6 per cent of the entire group. The interviews were conducted during the Winter and Spring of 2003. Among the interviewees were members from the seven most prominent parliamentary groups. Six of the interviewees had voted for the building of the fifth NPP unit; and an equal number had voted against it. Ten of the interviewees were chosen because of the positions they had adopted on the nuclear decision had remained uncertain for a long time or because their voting behaviour had differed from their earlier voting register on nuclear issues or from that of their reference groups. The hypothesis suggested that for these MPs, the decision was harder to make. This, in turn, would make them potentially more interesting in terms of understanding why the Parliament ratified the DiP in 2002. As a reference, two interviews with MPs who had clear profiles as long-time nuclear discussants were also conducted. One of these parliamentarians was a visible figure in the anti-nuclear movement while the other had long been fighting for the building of the fifth NPP unit.

A majority of the interviews took place face to face with the parliamentarians. The time lag of between eight and twelve months between the decision and the interviews appears to have had no noticeable negative impact on the quality of the answers. On the contrary, the answers seemed to have become riper with time and the MPs had forgotten many of their earlier slogans. Thus, the material now gave perhaps a clearer picture of parliamentarians' thoughts. In general, this opportunity to obtain a picture of the MPs' thoughts was also an advantage of the interview material over, for example, the plenary records: to know what was happening in the heads of the MPs was more valuable than what they intended to say to their colleagues in the plenary. Because the MPs perceived the decision on the fifth NPP unit to be an important one, they could discuss it with ease. Many held interesting perspectives on the nuclear decision and its meaning in a wider context, which made the interviews both rewarding and diverse.

The analysis of the interviews

The analysis of the interviews was a twofold process: On one hand, the data were treated as expert knowledge that reflected the ways the decision was made in the Parliament. On the other hand, the material was seen to contain interesting *discourses* on nuclear energy. Hajer (1995, 44, 60) defines discourses as different texts and ways of speaking in a certain context. Discourses can be analysed by considering the content, that is, the ideas, concepts and categories one can find in a speech or a text. On the other hand, it is also possible to identify the social practices behind the discourses by considering the context in which the speech or text exists. This analysis paid greater attention to the content, the ideas and categories and examined the systems of meaning and the competition between them.

The analysis revealed a wide variety of ways of discussing nuclear energy and the decision on the fifth NPP unit in Finland. Still, when the model with three ideal types of development (simple, ecological and reflexive modernization) was tested in conjunction with the material, the categories seemed able to reflect the most important dimensions of the discussion. The material from the interviews that dealt with the ideas about the role of expert and lay knowledge, economic growth and environmental risks seemed to fall rather neatly within the three categories. On the other hand, the analysis of the discussion on nuclear power revealed the need to split the category of ecological modernization into two.

Because the analysis of the interviews dealt partly with the testing of the theoretical framework, this choice clearly compromised the materially based nature of the study. On the other hand, this choice highlighted for the analysis a strong theoretical link to other discussions that deal with fundamental issues of development, technology, environment and power in our time. Another question focuses on what can still be said of the whole discursive field on the basis of the interviews with predominantly those MPs who had potentially struggled with their decision. This bias in the material potentially painted a clearer picture of the middle ground of the nuclear debate. This was interesting because the decision was not overwhelmingly in favour of the building of the fifth NPP unit: had eight MPs decided differently, the DiP would have been rejected.[1]

The characteristics of the discussion on the fifth NPP unit

> [I]t has been about being involved in a rather historical thing.
>
> Green parliamentarian who voted against the fifth NPP unit

The research material painted a broad picture of the characteristics of the nuclear discussion in the Finnish Parliament. One of the important dimensions was the fact that in the parliamentary groups of the political parties, the issue had been declared to be an *issue of conscience*. This meant that the political groups were not making common decisions, and that the parliamentarians had the freedom to vote for their preferred alternative. As Erika Säynässalo points out in this book, this practice has its historical roots in the strategic decision taken by the former PM Esko Aho in 1992. At that time, the Parliament took a position on nuclear energy in its reading of the government's account for the energy strategy of Finland. The prime minister noticed that the issue cut across party lines, so he gave the MPs the freedom to vote as they preferred. This practise has been retained ever since, which makes parliamentary decisions in relation to the nuclear power industry distinctive as well as interesting decision-making cases.

During the course of the interviews the special character of the nuclear discussion emerged clearly. The parliamentarians described the ambience of the process as being intense, emotional and even historical. The attention that the decision gained in the media and among voters, and the plurality of risks and possibilities attached to the decision, made the positions of the parliamentarians difficult. In particular, those MPs who were critical of nuclear energy also saw that the decision was linked to broader ideas about societal development and the responsibility towards future generations. The following quote illustrates a reflexive understanding of the nuclear decision.

> [N]uclear decision is a symbol of this kind of...big...could one say, not quite the choice of lifestyle. And as I said, my previous critical view has been based on just that I have seen the building of new nuclear power plants and the big use of nuclear energy as a symbolic message for the continuation of this kind of Western vanity lifestyle.
>
> Left Alliance parliamentarian who voted for the fifth NPP unit

On the other hand, many of the pro-nuclear parliamentarians saw that the reactions towards nuclear energy had been exaggerated and that even to grant the issue freedom of conscience status was unnecessary. Some felt that several groundless fears related to, for example, the final disposal of nuclear waste prevailed. These interviews reflected a willingness to *normalize* the question and to do away with the excess burden of meanings attached to the decision. In addition, the role of expert knowledge in making the decision was also emphasized:

> [T]his issue demands quite a lot of natural scientific and technical information and thus [i]t is a tricky thing to handle in the Parliament.
>
> National Coalition Party parliamentarian
> who voted for the fifth NPP unit

The fifth NPP unit and the discursive field

When analysing the nuclear discourses, the theoretical framework of simple, ecological and reflexive modernization was successfully tested as an organizing framework. It was easy to find discourses that would match the ideas of simple and reflexive modernity. The case of ecological modernization was more challenging as similar pro-environmental arguments that were also positive towards growth and technology were utilized both to defend and to defeat the building of the fifth NPP. The following sheds more light on the structure and special character of the different discourses.

The pro-nuclear progress discourse

The pro-nuclear progress discourse showed that the experiences of nuclear energy in Finland have been largely positive. In the discourse, the nuclear accident of Chernobyl was regarded as a special case that could not occur in the Finnish context. There was a strong trust in western – and particularly national – technology and engineering. Unfortunately, the interviews failed to reveal what defines the 'nationality' of the nuclear power since, for example, the supplier of Finland's fifth NPP unit is the Franco-German company Framatome ANP.

In the pro-nuclear progress discourse, the risks of nuclear energy were coped with by comparing them to other risks for safety and the environment such as terrorism, small particle emissions and climate change. The point was that, compared to many other risks of the modern world, the risks of nuclear accident and the final disposal of nuclear waste were seen to be strictly *under control*. While such risks could not be abolished, they could be defined as minimal in the Finnish case.

> In my opinion there are similar risks related to the use of nuclear energy as to any form of energy production. So, of course there are [risks], what possibilities have been discussed, if a nuclear accident or a terrorist attack occurs, but also what I believe that in Finland we have experience of nuclear energy and of the plants – and I feel that no other form of energy production do we follow this closely.
>
> Social Democrat parliamentarian who voted for the fifth NPP unit

One important aspect of the pro-nuclear progress discourse was the connection made between the increased use of nuclear energy and *economic growth*. A nuclear power plant was viewed as a substantial source of cheap electricity that would benefit both the households of the 'cold country' and also heavy industry. In addition, nuclear power promised to give Finland the opportunity to export energy. Because no clear definition of where and why Finland should export energy emerged in the course of the interviews, these wishes could perhaps be interpreted as general aspirations to end the country's energy dependency and replace it with 'Finnish' nuclear power.

In general, there was a strong tendency to think that the building of the fifth NPP unit would boost economic growth. However, talk about economic growth saw no emphasis on whether we should have more of something. On the contrary, such talk of economic growth stood for the continuation of present conditions. Thus, economic growth was understood in terms of stability rather than change. Moreover, maintaining the current *welfare state* could be considered to be a socially sustainable decision. The logic was that if we would continue the current course of energy policy and allow the industry to invest in the new NPP unit, Finland would have more resources to devote to health care and other social services. Such arguments seemed to give the nuclear alternative a softer focus. Under this discourse, the decision to build the fifth NPP unit stood for stability and social responsibility. This connection between the ideas of nuclear power and the welfare state has also been previously explored by Lammi (2004, 13–14).

> I think that the values at this point of the decision came from the whole – like why we need affordable electricity. And it creates that latitude, in my opinion, in the society to invest elsewhere, like there has now been talk about health care and many other things when society doesn't need to support the electricity production.
>
> Social Democrat parliamentarian who voted for the fifth NPP unit

The pro-nuclear progress discourse can be seen to represent the logics of *simple modernization*: The potential of the fifth NPP unit to produce large amounts of cheap energy is believed to outweigh the possible risks. Moreover, controlling such risks is considered the task of experts who have thus far taken good care of the issue, at least in Finland. An important consideration is that nuclear energy boosts economy, which in turn safeguards the future of the welfare state. The word 'progress' in the name of the discourse has been borrowed from a well-known article by Gamson and Modigliani (1989, 4), who used it to signify society's commitment to technological development and economic growth.

The pro-nuclear climate discourse

Interviews included a large amount of material indicating that environmental risks were also taken much more seriously than in the pro-nuclear progress discourse. On the other hand, a substantial amount of this material still highlighted standpoints that were familiar to the progress discourse, such as a belief in technology and the importance of economic growth. Thus, the underlying logics of this material can be seen to reflect the principles of ecological modernization. A distinct feature of the discourse presented here is the idea that nuclear energy is emissions-free. Therefore, it is called the pro-nuclear climate discourse.

In terms of discourse analysis, the idea that nuclear energy poses climate benefits is an interesting one: Instead of considering whether or not the risks of nuclear energy should be accepted, for example, for the sake of the economic good, one can consider whether the acceleration of climate change is a bigger threat than the problems related to nuclear energy. Moreover, as was pointed out above, as early as the 1990s, environmental opinion leaders considered climate change a greater global environmental threat than the use of nuclear energy (Järvelä and Wilenius 1996, 49). The comparison has made the risks of nuclear energy appear proportionate.

> [T]he nuclear power plants of this generation, as regards them, the risk is rather minimal that a nuclear accident would happen. So for me the bigger threat is the burning of the fossil fuels.
>
> Christian Democrat parliamentarian who voted against the fifth NPP unit

Climate change was not a novel political theme in 2002 when the decision on the fifth NPP unit was made in the Finnish Parliament. Its status on the political agenda, however, was considerably higher than had been the situation in 1993. In 1993 the Parliament of Finland rejected the building of a new NPP unit. On the other hand, since then, climate politics have leapfrogged and become economically relevant (Tirkkonen 2000). In general, the pro-nuclear decision of the year 2002 was a decent deal: The risks of climate change had made the threats of nuclear technology appear much less sinister. Meanwhile, the quest to cut CO_2 emissions had changed the image of nuclear technology from environmental enemy to actual *environmental remedy*. It had become 'cleaner' although, interestingly, not 'totally clean'. Throughout the process, nuclear technology had been able to retain its economic attractiveness and pro-welfare image. When it was further agreed that a vote for the fifth NPP would include additional economic support for renewable energy sources, the deal became almost irresistible. However, some MPs who now felt tempted to choose the nuclear alternative still wanted to limit its use. Because it was desired that the future would be 'totally clean', nuclear energy was seen as something that could be utilized while renewable energies were still in the development phase and economic growth was desired.

> [It should be] accepted that it is just about jumping over the pit so that when the new technology and better options make it possible, so we would also pledge to back off, even to back off in the long run [from using nuclear power].
>
> Left Alliance parliamentarian who voted for the fifth NPP unit

If we try to analyse the logic behind the pro-nuclear climate discourse, we can note that it really resembles the ideas of *ecological modernization*.

The attempt is to mitigate climate change by using technology and without drastically confusing the power relations among the experts or the course of economic growth. As Lampinen points out in this book, this combination of arguments closely resembles the principal set of arguments presented by TVO in its application for the DiP. Thus, we can think that either the TVO was well aware of the political currents of the time and chose its words wisely or, alternatively, that it was particularly efficient in conveying its message. Moreover, both of these reasons may have affected the similarities. In any event, the pro-nuclear climate discourse provides for those who identify themselves as environmentally minded to defend the status quo. On the other hand, it could offer an easy way for the more economically oriented MPs to familiarize themselves with environmental argumentation. Thus, it is interesting to see what this new compromise field will yield in future discussions on energy and the environment.

The pro-renewables climate discourse

The third discourse in the discussion on the fifth NPP unit was in many ways based on similar assumptions to the pro-nuclear climate discourse: both economic growth and technological development were emphasized alongside the environmental arguments. Within this discourse, however, some of the claimed benefits of nuclear energy were strongly questioned. Is nuclear power really an economically viable investment? How do the Finnish plans to build the fifth NPP appear from an international perspective? Why invest in nuclear technology when the production of electricity is possible with fewer environmental risks? This discourse, which is labelled the pro-renewables climate discourse, sees nuclear power as old-fashioned, potentially expensive and problematic for society as a whole with different ramifications. It does not aim to abide by the use of possibly 'cleaner' nuclear power, but finds it possible to choose the *'cleanest'* alternative right away.

> [N]uclear power has been lobbied for almost thirty years so certainly it has decelerated all other innovations, at least their financing, so that it has not been worthwhile to invest as it has been uncertain how we are going to satisfy our energy needs.
>
> National Coalition Party parliamentarian who voted against the fifth NPP unit

> [W]hen nuclear power plants are built, there is the need to build so much standby power capacity by its side – in that a lot of coal is used that then pollutes the environment.
>
> Social Democrat parliamentarian who voted against the fifth NPP unit

Referring back to the ideal models of development, the pro-renewables climate discourse is a textbook example of the logic of *ecological modernization*. Alongside expert knowledge and modern technology, it emphasizes the structural change of energy production towards renewable energy forms. The pro-renewables climate discourse appears to be the closest representative of the message of the environmental NGOs in the 2002 debate. As Lammi points out in this book, the language of the NGOs was focused on technology and economics. In general, the decisive combats in the nuclear discussion of 2000–02 apparently took place within the compromise field of broadly defined ecological modernization. The combats represented struggles between two discourses, the pro- and anti-nuclear climate discourses, one of which was already proposed in the TVO application while the other was adopted by the environmental NGOs. If so, it was no longer necessary to discuss whether we should take environmental problems seriously, whether the economy should continue growing and whether modern technology should be utilized in solving the problems as these were already taken for granted by both of the important camps.

The reflexive anti-nuclear discourse

> In the over-generational perspective, we never know what catastrophes may follow if emissions or accidents, terrorist attacks occur. And in general, when things are examined from the perspective of thousands or even hundreds of years, there are unknown risks that, if realised, may be catastrophic. These questions are then ethical by nature, when generations and things you cannot be responsible for are in question.
>
> Left Alliance parliamentarian who voted against the fifth NPP unit

According to my material, the most common way to criticize nuclear power was to cite an accident, such as that of Chernobyl, or problems related to the final disposal of nuclear waste. When discussing the risks, the question arises of whether or not they outweigh the benefits. Moreover, Beck (1992) also asks whether the risks are *acceptable*. And can we count on the underlying logic of such a society that produces self-endangering risks? This is an important aspect in the reflexive anti-nuclear discourse presented as the last, but not the least, of the discourses in the 2002 nuclear debate. This discourse finds the risks of nuclear energy to be unbearable and it questions the ability of technology and expertise to solve them. Furthermore, it sees the environmental risks we face as indicators of the problematic direction of societal development. Thus, tackling them would demand a thorough change in both attitudes and policies. Here, the idea of constant (economic) growth is among the ideas that are most often criticized.

> [I]n that sense, it is a question of values whether we need to follow the path of constant growth in all our consumption. Is that the right direction

> for the development of the whole world? Every day we consume more of this Tellus of ours, so that for me there would be the maximum amount of all good and as easily as possible. It is deeply a question of values.
>
> Christian Democrat parliamentarian who voted against the fifth NPP unit

The reflexive anti-nuclear discourse apparently echoes the principles of reflexive modernization. While this discourse concentrated more on the risk profile of nuclear energy, it appears to reflect the same ideas as the messages of the environmental NGOs in Finland's 1993 nuclear debate. As Lammi writes in this book, at that time, nuclear safety and nuclear waste were among the most central themes to be taken up by the environmental lobby. Contrary to the nuclear votes of the new millennium, however, the anti-nuclear alliance won the case.

One interesting feature of reflexive modernization is that it recognises nuclear power to be a politically relevant *symbol*. It opens a discussion about different symbolical dimensions attached to nuclear power. What does it stand for? What does it signify? It notes that energy production is a very important part of modern society. Electricity not only keeps the plants running, but also lights and heats our homes, keeps information circulating and provides the power for our household appliances. Electricity and energy in general have become such basic commodities in our society that it is easy to think that their potential to resonate is strong. In addition, nuclear energy has gained special resonance as a controversial environmental issue. Even though pondering about the symbolic meaning of nuclear energy is a reflexive act in itself, one must note that that the nuclear energy industry means different things to different actors. As has already been shown in our exploration of the different discourses, nuclear energy is to some a symbol of wealth and welfare state, while to others it is a signal of old-fashioned technology and unnecessary risk. Still, the mere recognition that a certain form of energy could hold strong cultural meanings and effects shakes the power position of simple modernisation in particular.

The four discursive dimensions and the fifth NPP unit

Table 4.1 provides a rough summary of the discourses of the fifth NPP unit. The summary is based on the three different ideal types of development presented earlier in this chapter. Therefore, the table emphasizes those aspects of the discourses that are important in distinguishing between ideal types of development. These aspects deal with views on nuclear energy, environmental risks and economic growth.

The model shows how the compromise field of ecological modernization functions between the contradicting discourses of simple and reflexive modernization. It visualizes the situation in which the existence and importance of environmental problems is recognised in wider circles. The model indicates

Table 4.1 The discourses of the discussion on the fifth NPP unit

Discourses	**The pro-nuclear progress discourse**	**The pro-nuclear climate discourse**	**The pro-renewables climate discourse**	**The reflexive anti-nuclear discourse**
The ideal types of development	Simple modernization	Ecological modernization	Ecological modernization	Reflexive modernization
View of nuclear energy	Nuclear power enables us to produce enough energy at an affordable price.	Nuclear power is an efficient way of cutting the greenhouse gas (GHG) emissions.	Nuclear power is an old-fashioned and expensive way of producing energy and carries unnecessary risks.	The environmental risks of nuclear energy are excessive. Its whole way of producing energy reflects the logics of modern self-endangerment, and thus must be overturned.
View of environmental risks	Environmental risks require no structural adjustments, but end-of-the-pipe solutions can sufficiently eliminate them.	Better control of environmental risks demands structural adjustments that cut the GHG emissions of the energy sector, for instance, by building a new NPP. Institutionalized experts play key roles in generating solutions.	Better control of environmental risks demands structural adjustments; the energy sector also requires fundamental change towards the use of renewable energy sources. A wider array of experts play key roles in generating solutions.	The existence of environmental problems is an indicator of the problematic nature of current societal development. Changing the course requires a throughout change of thinking that involves values, emotions and counter-expertise.
View on economic growth	Economic growth permits the longevity of the welfare state.	Economic growth permits the longevity of the welfare state, but growth must be 'cleaner'.	Economic growth permits the longevity of the welfare state, but growth must be 'clean'.	The idea of constant growth is contrary to the idea of sustainable development.

that while there is widespread support for reacting to environmental risks, thoughts critical of the modern growth logics have not been adopted in broader circles. The model attempts to *reformulate the bipolar ways of framing the nuclear discussion*. In addition, it sheds more light on potential frontiers where discursive battles may take place. How is it related to the previous research in the field?

Koskinen (1995, 337–43) has analysed the discussion on Finland's nuclear energy law during the years 1985–86. He found in the discussion a division between productionist and ecological worldviews. Koskinen's classification has much in common with that of Litmanen (2004, 223–5), where the worldviews presented in the nuclear discussion of 1993 are called industrial and conservationist. As regards the Finnish nuclear discussion of 2000–02, many have pointed out that it changed the old polarizations. Litmanen (ibid.), for example, came upon that the deep structures of nuclear discourses have been shifting, and that the old frontiers have been unravelling. On the other hand, Lammi (2004, 46) found in the discussion a new polarization between two distinct stories: In one of the stories, nuclear energy stands for employment and the welfare state. In another, nuclear energy represents the deepening of the unsustainable western industrial policy twist. The analysis presented here is not contradictory but complementary to the results of Litmanen and Lammi. The stories Lammi found in relation to nuclear energy seem closely to resemble the discursive extremes of my model: the pro-nuclear progress discourse and the reflexive anti-nuclear discourse. Meanwhile, the middle ground of my model, where the ideas of ecological modernization prevail, concretizes Litmanen's notion that the deep structures of the nuclear discussion are changing.

Still, it needs to be noted that Nurro (1995) had already found an environmentally grounded nuclear discourse in the Finnish energy discussion of 1985–93. Nurro's research material included periodical magazines and interviews chosen to reflect the views of the 'nuclear technocracy'. In this context, the environmental discourse emphasized Finland's global responsibility, the safety of western and particularly 'Finnish' nuclear power, and the capacity of nuclear energy to cut greenhouse gases. Nurro (ibid., 66) points out that in her material at least, the growth discourse, the idea that more cheap electricity is needed, was a more fundamental idea than the environmental friendliness of the energy sector. This view is further shared by Suotula (2000, 67–9), who found that ecological arguments were used 71 times in the 1993 parliamentary discussion on nuclear power. Suotula notes that almost half (46 per cent) of these arguments were used by the National Coalition Party, which has no strong tradition of taking pro-environment stands in political discussions. Thus, Suotula's 'basic power discourse', which emphasizes economic rationalities such as growth and employment, included pro-environmental nuclear discourses (Suotola 2000, 117–20). This could be interpreted to mean that Suotula sees the pro-nuclear environmental argumentation of the 1993 debate as superficial rhetoric. Thus, her results do not conflict with the categorization of Litmanen (2004) (presented above), who labelled the worldviews expressed in the 1993 nuclear discussion as industrial and conservationist.

One critical question that remains unanswered is whether the pro-nuclear climate discourse of the 2002 debate was more 'greenwash' than a sign of 'real' green thinking. Here, the evidence suggests genuine environmental concern was involved in the pro-nuclear climate discourse of the parliamentarians, at least among those interviewees who were neither strongly pro- nor anti-nuclear. Even though the pro-nuclear discourses of my material reflected perhaps more clearly the social and economic concerns than the anti-nuclear ones, it would be unfair to say that growth discourse was their most fundamental characteristic. Thus, the pro-environmental nuclear discourse may have grown in importance hand in hand with the ideas of ecological modernization both by spreading across society and by deepening in meaning. This would fit with the results of Suotula (2000, 128), who sees that the rise of environmental problems is something that has strengthened the possibility of achieving consensus on energy questions in the Parliament of Finland. Although the division between MPs representing technical and value-rational standpoints was still quite clear even in the 1990s, Suotula argues that environmental concern is a question that wins support across party lines and that can thus function as a mediating theme.

In short, environmental questions, and in particular climate change, clearly had a new role to play in the 2000–02 nuclear decision-making process of the Finnish Parliament. Nuclear energy was no longer being defended simply by 'hard' economic argumentation and opposed by 'soft' environmental language. Rather, the decisive battle over meaning now seemed to take place in the middle ground where the logics of ecological modernization could serve both for defending and opposing the building of the new NPP unit. Environmental questions, and particularly climate change, played prominent roles in the debate. At the same time, however, environmental NGOs had to argue within a discourse that was not only *unfavourable for their counter-expertise*, but instead gave preference to the representatives of institutionally strong expert positions. The next section will shed more light on the way the interviewed MPs perceived different actors.

The role of actors in the decision on the fifth NPP unit

> [T]he decision of the Parliament is based on the fact that many committees deal with issues, and in that way parliamentarians obtain information about different kinds of aspects. Ordinary people don't

> have that knowledge and background, so they are in a rather difficult situation.
>
> National Coalition Party parliamentarian who voted for the fifth NPP unit

The section above analysed the ways the parliamentarians saw the discussion on the fifth NPP unit. What was the decision all about? This section deals with the central decision-making actors as seen by the MPs. The analysis concentrates on three different actor groups in particular: citizens, experts and the parliamentarians themselves. In any way, all the actors that were authorities of some of the central themes of the nuclear discussion could be considered experts. As there was no consensus in the Parliament on which themes were central and which were not, the question of expertise was debatable. Who was allowed in the exclusive circle of experts? And who was excluded? And how could the insights and information the experts provided be used for shaping and supporting the standpoint one was about to take?

As regards the relationship between citizens and parliamentarians, the Constitution of Finland (Chapter 1, Section 2) clearly states that 'The powers of the State in Finland are vested in the people, who are represented by the Parliament'. Still, it was interesting to hear how the parliamentarians saw their potential voters, the role of public opinion and of NGOs. What kind of role should the citizen's voice play in a discussion about building a new nuclear power plant? Could the question of the fifth NPP unit have been put to a referendum?

Parliamentarians as decision makers

The interviews with the parliamentarians focused in depth on the decision-making process of the new NPP unit. In the discussions, some of the most common themes that arose were the mandate given by the people and the responsibility that parliamentarians as decision makers had to bear. The obligations of the parliamentarians could be defined in at least two ways: On the one hand, the MPs felt they should act as they had *promised* before the elections. On the other hand, parliamentarians also saw that they were responsible not only to their voters, but to *society as a whole*. One should consider the common good when making such a decision. With regard to responsibility, however, some parliamentarians found it hard to bear in this case. They saw that a new nuclear power plant would have such far-reaching effects that it would go beyond the decision-making powers of today's parliamentarians:

> [Y]ou should make decisions for which you can bear the responsibility and you should avoid decisions for which you cannot bear the responsibility. Nuclear power is the kind of thing for which bearing the responsibility in

> the long run is impossible for today's decision-makers. In that, you shift the responsibility and the price to future generations.
>
> Christian Democrat parliamentarian who voted against the fifth NPP unit

In the interviews, many of the parliamentarians acknowledged the importance of the *parliamentary reference groups* in everyday decision making. Now that nuclear power was declared a question for which the freedom of conscience prevailed, parliamentarians were left alone as decision makers. However, even though it was no longer obligatory to vote along party lines, loyalty to one's background groups still influenced the voting behaviour. The nuclear decision could, for example, be a conscious or unconscious way for opposition MPs to 'revolt' against the government's proposal to make the decision to build the new unit. As the result of the vote was less predictable than usual, the decision-making process activated the opposition in general. In addition to political groups that were either inside or outside the government, the parliamentarians also belonged to certain committees. The Parliament of Finland currently has 16 committees, and the hearings, discussions and excursions they organize form an integral part of the Parliament's functioning. The committees were perceived by some MPs as important reference groups. On the other hand, many also viewed them as privileged opportunities to obtain knowledge:

> [T]he committee work of the Parliament is excellent in that you can invite the expert you want. So if you are interested in health issues or if you want to know about the storage thing as regards nuclear waste, so you get [information] directly from the state apparatus, from the Radiation Safety Centre, you get engineering offices, you get what you want. So you have an opportunity that doesn't really exist elsewhere.
>
> Left Alliance parliamentarian who voted against the fifth NPP

Broad knowledge was defined in many interviews as the basis for good decisions. Many felt they had learned a lot during the decision-making process. On the other hand, MPs' privileged *access to knowledge* also seemed to give the MPs a cause to lift themselves above the other lay people of energy policy. Because ordinary interested citizens had no similar access to knowledge, their opinions carried less currency than did those of the parliamentarians. On the other hand, some parliamentarians were critical of the amount of information being poured into the process. Was it really possible to digest it all? Or was it just about giving political power into the hands of experts? In this situation, some MPs decided to ignore the details and preferred broader principles. Holding views was considered the most important thing in politics. However, a view differed from a mere opinion or feeling as it was something that developed during the decision-making process of the fifth NPP unit. A view was constructed while being exposed to a wide range of information.

It was affected by discussions with different people as well as political experiences. It helped in the active gathering of information: What is important and what is irrelevant? Whom should one believe?

> [I]t's like in politics in general that how much information you have is not necessarily important – exact knowledge. In politics, the overall view counts; what kind of view you have on the whole. It is the same thing in this nuclear power case, I guess and the view, then, is certainly comprised of one's own values, morals, ideology and the knowledge one receives.
>
> Left Alliance parliamentarian who voted for the fifth NPP

To sum up, parliamentarians found support for their decision-making process from at least three sources: from the expert hearings in the committees, from their other background groups such as parliamentary and interest groups, and from their own broader views that had developed over time. Even though many of the parliamentarians seemed well aware of their elevated position in the Finnish political system, they were unafraid to reveal the human face of decision making. The limitations of handling information were acknowledged as well as the emotions and power struggles attached to the decision:

> It was a very difficult situation. We parliamentarians need to listen and take an elementary course in energy policy, nuclear energy – so suddenly we have experts that end up having totally contradictory conclusions, so it's very hard to figure out.
>
> Social Democrat parliamentarian who voted against the fifth NPP unit

> Well, I would say that in '93 there was more drive in the Parliament to oppose nuclear power, – and now, there was somehow the feeling that in the Parliament there was, on the contrary, enough willingness to find the yes arguments. So, I would say that both decisions were based on emotion, but what caused that emotion [is the question].
>
> Green parliamentarian who voted against the fifth NPP unit

The role of experts in the decision-making process

In the interviews conducted with the MPs, the question about the role of citizens and experts was dealt with, but not in detail. As many other authors have also touched upon the issue in the context of the 2000–02 nuclear discussion in Finland, the results in this study concur largely with the findings of others. Therefore, the following deals not only with this analysis, but also with those of others. As noted previously, *experts* were important actors in the decision-making process of the MPs. According to Hylkilä (2003, 136–8), in the discussion on the fifth NPP, parliamentarians relied heavily on experts with a strong institutional background, such as university professors and ministry representatives; my own interviews confirmed this finding. In addition, the interviews attested the phenomenon, which Litmanen covers in Chapter 7, that Finns in general count on expertise, technology and authorities.

In politics, *interest groups* are often enjoying a rather broad expert status (Ruostetsaari 1987). In addition, certain interest groups are typically close to certain political groups and parties. In the debate on the new unit, interest groups played vital roles. Many of the biggest and most influential groups, such as the Central Union of Agricultural Producers and Forest Owners (MTK), the Central Organisation of Finnish Trade Unions (SAK) and the Confederation of Finnish Industry and Employers (TT), were clearly in favour of building of more nuclear power. According to Hylkilä (2003, 136–8), these expressions of support strongly affected the MPs in parties that were close to these groups. The information provided by the interest organizations was considered less reliable than that delivered by independent experts, such as professors and ministry representatives. Still, the opinions of the groups were considered important due to their representative function in important social sectors:

> [I]f I now need to name something, if only because of my background, I surely listened carefully to the views of the trade unions, the views of the industry.
>
> Left Alliance parliamentarian who voted for the fifth NPP unit

Lammi (2004, 30) has noted that in the debate about nuclear power, both decision makers and the media felt that *who* was talking was more important than *what* was being said, for example, about the economy. This study further reaffirmed this view as the parliamentarians emphasized the background of the experts. On the other hand, the parliamentarians also valued certain ways of presenting the information. Many of the interviewees said that the lobbying work and argumentation during the 2000–02 nuclear discussion was calm and well informed. This notion supports Kojo's (2004, 239–42) point that the pro-nuclear alliance in particular had improved its strategy since the 1993 nuclear vote, when the Parliament of Finland rejected plans to build more nuclear power plants.

When defining what experts are, it is equally important to know who is *not* considered an expert. In this context, one central objective was to determine what constituted a biased or prejudiced attitude. If an address was considered biased, it was no longer considered competent. Here, the environmental organizations in particular were targets of criticism.

> I think that the credibility was still higher for those who could argue both sides [of the issue] and understand the other opinions. So let's say in the

nature protection organisations – [W]hen they visited our Finance Committee or other committees, if it was just one-sided, it provoked a kind of counter-reaction. So in one's heart of hearts, people were laughing that it is not quite like that.

Left Alliance parliamentarian who voted for the fifth NPP unit

It is interesting that only environmental organizations were spontaneously labelled as biased as their strategy was to argue in favour of what they considered technically superb energy solutions and ways in which nuclear energy would affect the economy (see Chapter 3 in this book). It seems, as Hylkilä (2003, 116) has pointed out, that the anti-nuclear alliance had a problem with its profile. For many parliamentarians, opponents of the new NPP unit remained a vague, unfocused group with which they had no close interaction. Moreover, it is worthwhile to consider whether the criteria for expertise were narrower in this discussion than in the 1993 nuclear debate. If a calm and well-informed civil servant was the ideal type of expert in this case, surely the environmental NGOs found it difficult to make their representatives fit this image. In addition, they may also have lost their status as an important interest group now that the environment was seen as a concern for all. The pro-nuclear lobby had taken the environmental argument and packaged it neatly together with the fifth NPP proposal, but, unlike many others, the environmental NGOs were unwilling to accept it. Was this the reason why the anti-nuclear NGOs were considered to be biased, while the pro-nuclear industry was viewed as balanced?

The divided picture over citizen participation

A study by Kiljunen (2007) reveals that 44 per cent of Finns are concerned about the unknown hazards related to nuclear power. An overwhelming majority of respondents (82 per cent) assume that if there were an accident, it would inevitably cause irreversible damage to extensive areas and large numbers of people. During the interviews, many interesting discussions touched on the role of ordinary citizens in the nuclear debates and decisions. Some of the clearest of these discussions dealt with the question of whether a *referendum* should have been organized on the fifth NPP unit. According to the Constitution of Finland (Chapter 4, Section 53), it is possible to organize consultative referendums that would not be legally binding on the Parliament, but that could, of course, easily be interpreted as politically binding. Some of the interviewees accepted the idea that public opinion should play a more direct role in decision making on nuclear power and that a referendum on the new unit would have been a good idea. The point was that because the nuclear decision would include big risks and have far-reaching consequences, it would be a worthwhile exercise to ask whether or not the people were ready to take the risk. Thus, some also considered emotions to be a legitimate basis for a decision. In the end, it was considered important that people have the opportunity to participate if they so desire.

> [N]uclear power could be the kind of thing on which you could well have a referendum, since it also deals with emotions, values you need to give them a say, too. So if people have phobias or fears as regards the nuclear power plant, then the question is not about whether the energy is cheap or expensive.
>
> Social Democrat parliamentarian who voted against the fifth NPP

If some of the parliamentarians found *emotions* to constitute legitimate grounds for decision making, many others openly despised the idea. For these MPs, the nuclear decision was a technical solution to the energy needs and climate commitments of the country. In addition, the majority of the parliamentarians interviewed were afraid that people would make a *poorly informed* decision in the case of the fifth NPP unit. The decision had been difficult and, in addition, the amount of information confusing for the parliamentarians. How would the information reach the people? Moreover, the question arose of whether or not the people would bother to familiarize themselves with the information they received. And even if they could carefully acquaint themselves with the nuclear knowledge they would rely on the biased media as their main source of information:

> The amount of information that exists on this issue cannot reach all the people in a centralised manner; otherwise, we would have to watch documentary films on the issue for a year or two, which wouldn't even interest people much, so in my opinion this is not an issue on which we would need a referendum.
>
> Christian Democrat parliamentarian who voted against the fifth NPP

On the other hand, some parliamentarians found the nuclear decision not to be too difficult for the people to make, but nevertheless preferred that the Parliament should make the decision. One reason for this was that representative democracy in general was considered to be a well-functioning system. In addition, MPs pointed out that decision making was the very reason for their presence in parliament. Once one had obtained the mandate to make decisions on behalf of others, it was considered poor form to 'run behind the backs of the people' when a hard decision loomed:

> [S]urely people had an opinion on it, like when the Gallup polls were made, so quite many could and wanted to speak out. And I think that the knowledge of the ordinary man is sufficient; there is no need for any top expertise in making decisions on these issues. So if it were like that,

there would be no need for parliamentarians or others; just mathematical equations and that's it.

(Interviewer) So, would there have been a need to organise a referendum on this?

No.

(Interviewer) Umm... Not of any kind?

In my opinion there was no need.

Centre Party parliamentarian who voted against the fifth NPP unit

Hylkilä (2003, 122) concludes that the citizens' lobbying of parliamentarians about the fifth NPP unit had little effect. This once again affirms the notion that citizens have rather limited power in questions that deal with scientifically defined environmental risks. Moreover, it seems that the parliamentarians' image of a citizen still seems to resemble 'the man in the street' or 'the girl next door', even though the Finnish population today is well educated and many people may indeed have interesting things to say about nuclear power on the basis of their working life or leisure activities.

(Interviewer) How easy do you think it would have been for an ordinary citizen to make this decision?

It is easy when you don't have to make it, when you can say it there – in a market place or in a pub – that I'm for, I'm against. But here in the Parliament, it was of course a pretty tough decision.

National Coalition Party parliamentarian
who voted against the fifth NPP unit

The three theoretically compatible discourses on actors

Once again, what influenced the parliamentarians' perceptions of the actors was the general nuclear discourse or discourses they were using. When the debate on the new unit took place, competition erupted both among the different nuclear discourses as well as among the players. Interplay also emerged between these two domains as certain discourses strengthen the role of certain players, and the players actively attempt to shape the discourses. Moreover, it is also possible to construct discourses about MPs' views of actors. This is what appears in Table 4.2, which summarizes some of the most important findings. As the aim of this chapter was to use the same theoretical basis (the division between simple, ecological and reflexive modernization) also for the material on actors, the discourses presented do not cover the whole richness of the material. This is also the reason why the plurality of the MPs' views on each actor group was presented first, and only now the discourses.

Still, the identification of these discourses was rather easy and the model seems to reflect some of the most theoretically relevant notions as regards parliamentarians, experts and citizens in nuclear discussions: The 'powerful

Table 4.2 The discourses on the roles of different actors in the discussion on the fifth NPP unit

Discourses	**Powerful experts and uninformed citizens**	**Broader expert circles and citizen participation**	**Criticism of expertise and empowerment of citizens**
The ideal types of development	Simple modernization	Ecological modernization	Reflexive modernization
View on the role of MPs	A parliamentarian should not escape his/her responsibility behind the backs of the citizens. MPs are in a privileged position to obtain knowledge on issues.	A parliamentarian needs knowledge from a wide circle of experts.	A parliamentarian should not escape his/her responsibility behind the backs of the experts. The political view one holds should be utilized.
View on experts	Expert knowledge is a prerequisite for a good decision on nuclear energy. Western, and particularly Finnish, expertise guarantees the safety of nuclear technology. Moreover, powerful interest groups are worth listening to, but are less trustworthy than independent experts. That the lobbyists have become better informed and less bigoted is a positive thing.	The fact that the experts' knowledge often conflicts makes decision-making difficult.	The Finnish trust in authorities is problematic. The field of experts' knowledge is often narrow. One must trust one's own values and views in making such a decision.
View on citizens	Citizens have insufficient knowledge and time to familiarize themselves with issues related to nuclear energy. A referendum on the fifth NPP would have been a bad idea as citizens would have made their decisions on a weak basis.	It is important to involve citizens in a decision that piques wide interest and that affects the people.	You do not need top expertise to make a decision on nuclear power; knowledge within the reach of citizens, their values and views are sufficient. Citizens should be involved in making decisions that have far-reaching consequences and that bear risks for catastrophic accidents.

experts and uninformed citizens' discourse reflects the principles of simple modernization whereas the 'criticism of expertise and the empowerment of citizens' discourse could be considered to be representative of the ideas of reflexive modernization. It should be noted that in this model, the 'broader expert circles and citizen participation' discourse, which represents the thinking of ecological modernization, has more in common with the 'powerful experts and uninformed citizens' discourse than with the 'criticism of expertise and empowerment of citizens' discourse. The reason is that ecological modernization is not as radical in making the power shift as reflexive modernization. Ecological modernization is concerned with continuing the same tradition of Enlightenment as simple modernization, which means an emphasis on expertise and technology. Still, it leaves room for broader expert circles, tries to cope with conflicting information, and finds citizen participation to be important, at least in principle.

In general, in the debate on the fifth NPP unit, many factors have emphasized the role of individual decision-making processes and *trust* in certain actors as sources of information and insight: First of all, the important role parties and parliamentary groups usually play in opinion formation was diminished due to the declaration of the fifth NPP unit as an issue of conscience. Secondly, plenty of expert information was available on the topic, part of which was contradictory and confusing to the parliamentarians. Thirdly, the discursive field was not set, but changing. As the situation turned out in the 2000–02 nuclear discussion, the power positions of the citizens and counter-experts were somewhat weaker than those of the established experts. Little room was available for rebellion or distrust of the authorities. Still, many citizens rallied around the issue and the demonstration before the nuclear vote in the spring of 2002, for example, was the largest environmental demonstration in the history of Finland (Nissinen 2004, 99). However, many parliamentarians felt that people in general had little interest in the case and, even if they were interested, their likelihood of making well-informed decisions was limited. This legitimized their abandonment of, for example, the concept of consultative referendum on the fifth NPP unit.

Conclusions: nuclear power – from enemy to remedy

In summary, the 2000–02 debate about the fifth NPP unit in the Parliament of Finland had several distinctive characteristics. The process raised the political temperature and captured a great deal of public attention. The fact that the question had been declared an issue of conscience, the abundance and contradictory nature of expert information, and the changing face of the discursive field made it a difficult decision for many parliamentarians. Moreover, those MPs in particular who were critical of nuclear energy also saw that the decision was linked to broader ideas of societal development. At the same time, however, there were also tendencies and a willingness to normalize the nuclear issue. This was true particularly among those parliamentarians who viewed the fifth NPP unit as a technical solution for Finland's energy needs as well as for its climate commitments.

Previous studies (e.g. Koskinen 1995; Lammi 2004; Litmanen 2004) indicate how Finnish nuclear decisions have raised conflicting worldviews in the field. Sjölander (2004) has also shown how a nuclear waste management case in Sweden embraced a complex web of pressing issues for today's societies. Thus, the section on theory began with some core elements and presented the three ideal types of development: simple, ecological and reflexive modernization. The hypothesis predicted that a new discursive middle ground had partly reconciled the old contradiction between simple and reflexive modernization. The question focused on whether or not the logic of ecological modernization was capable of fostering the compromise. Moreover, the idea was that a deeper understanding of the ways in which nuclear energy was discussed would help in analysing the decisions made by the parliamentarians.

In this study of focused interviews conducted with the MPs, four general nuclear discourses emerged: (1) The pro-nuclear progress discourse represented the logic of simple modernization with an emphasis on economic growth, established experts and the end-of-the-pipe style of controlling environmental risks. Meanwhile, the other pro-nuclear discourse, (2) the pro-nuclear climate discourse, appeared to reflect the principles of ecological modernization. In that, economic growth and trust in modern technology and expert knowledge were important principles, but environmental risks were also taken much more seriously.

Another discourse that functioned according to the logic of ecological modernization was (3) the pro-renewables climate discourse. Alongside technological innovation and expert knowledge, this emphasized structural changes in the energy sector, for example, and the utilization of renewable energy sources instead of nuclear power. Thus, it represented the other anti-nuclear discourse alongside (4) the reflexive anti-nuclear discourse, which differed from all of the other discourses in its criticism of the chances of growth and expertise to solve environmental risks caused by modernity. Moreover, the reflexive anti-nuclear discourse found the risks of nuclear energy to be unbearable. They were considered to be indicators of the problematic direction of social development.

These findings suggest that the bipolar ways of seeing the nuclear discussion no longer reflect the array of nuclear discourses in Finland. In addition, the results also point out that there may have been a shift of the frontiers where combats over environmental risks now take place. While environmental questions have enjoyed wider recognition, the risks no longer challenge society as a whole: The ideals of ecological modernization have allowed power relations to remain the same, thereby creating difficult times for less powerful actors, such as environmental NGOs. Economic growth, institutional expertise and modern technology are seen as means to solve, rather than cause, environmental problems. Meanwhile, climate change and the environmental risks of nuclear power have been adopted as risks that society must handle. Climate change serves to downplay the risks of nuclear energy, and nuclear energy to mitigate climate change.

As regards actors, parliamentarians found support for their decision-making process from at least three sources: the expert hearings in the committees; their background groups, such as parliamentary groups; and from their own broader political views that they had developed over time. Thus, parliamentary practices played a crucial role in their decision-making process while the media and citizen contacts played a lesser role. Moreover, parliamentarians relied heavily on experts with strong institutional backgrounds. In addition, the views of the interest groups were also considered to be worthwhile, particularly by those MPs in close relationships with those groups. The information they provided, however, was considered less reliable than the information delivered by the independent experts.

If the presentations in the parliamentary hearings were capable of being interpreted as biased, the expert status could be lost. This happened, for example, to environmental organizations. Even though their strategy was to concentrate on the economy, technology and scientific facts, some parliamentarians spontaneously perceived the environmental NGOs as biased and prejudiced. An interesting question is whether this idea of biased environmentalists resulted from the environmental NGOs' refusal to support the pro-nuclear climate discourse. As some MPs and many other influential actors adopted and promoted this discourse, the position of the environmental NGOs must have appeared biased. In addition, wider circles have since adopted the position of environmental concern. Did this mean that the environmental NGOs had lost their special status as 'representatives for the environment'? Was it a strategy for some to make the environmental groups seem needless? And why were the industry representatives more successful at becoming the new envoys of the environment than the environmental NGOs were at talking about the economy and employment?

The opinions of the MPs were divided when it came to the question of the extent to which citizens should play a role in the decision making about nuclear energy. Some parliamentarians thought that a referendum on the fifth NPP unit, for example, could have been a good idea as the decision affects many people, involves considerable risks and has far-reaching consequences. However, many of the parliamentarians interviewed feared that people would make poorly informed decisions. The decision-making process had been difficult, and the amount of information confusing for the parliamentarians. How would the citizens have coped with the challenge? Many MPs also emphasized that they believed in the representative system and, thus, parliamentarians should make the decisions they had been elected to make.

To conclude, on the basis of the analysis, it seems that a shift has occurred in the way that nuclear energy is handled in Finland. After the Chernobyl accident in 1986, ideas about risk society and reflexive modernization appear to have gained currency and popularity that were still running strong in the beginning of the 1990s. However, as time passed, the situation normalized and the wider temporary distrust in modern technology and expert knowledge gradually evolved into an optimistic ecological modernization that better suits the national spirit in Finland. Over the past 15 years at least, the politics of climate change have grown steadily in importance. As nuclear power has been successfully defined as emission-free, choosing the nuclear alternative as a 'cleaner' energy source has become possible. In the decision on the fifth NPP unit, the nuclear alternative also provided a decent deal in which parliamentarians could vote for both the mitigation of climate change and the defence of economic growth as well as for the environmentally and socially responsible welfare state.

Notes

1. The vote on the ratification of the DiP to build the fifth NPP unit took place in the Parliament of Finland on 24 May 2002. The result was 107–92 in favour of building the reactor. Of the parties, the MPs of the National Coalition Party were largely in favour of the decision. Meanwhile, all of the parliamentarians of the Green League opposed the building of a new NPP unit. In addition, more anti-nuclear voters could be found in the opposition parties and among the female parliamentarians than in the government parties and among the male MPs.

References

Beck, U. (1992) *Risk Society: Towards a New Modernity*. London, Newbury Park and New Delhi: Sage Publications.

Beck, U. (1994) The Reinvention of Politics. Towards a Theory of Reflexive Modernization. In U. Beck, A. Giddens and S. Lash (1994) *Reflexive Modernization: Politics, Tradition and Aesthetics in the Modern Social Order*. Cambridge: Polity Press, pp. 1–55.

Beck, U., A. Giddens and S. Lash (1994) *Reflexive Modernization: Politics, Tradition and Aesthetics in the Modern Social Order*. Cambridge: Polity Press.

Berg, A. (2004) *Ydinvoiman uudet vaatteet. Kansanedustajan näkökulma vuosien 2000–2002 ydinvoimakeskusteluun* (The New Clothes of Nuclear Power. Parliamentarian's viewpoint to the nuclear discussion of 2000–2002). Unpublished master thesis of journalism and mass communication. University of Tampere, Tampere, http://tutkielmat.uta.fi/pdf/gradu00424.pdf – accessed 23 January 2009.

Constitution of Finland (1999) http://www.finlex.fi/fi/laki/kaannokset/1999/en19990731.pdf – accessed 23 January 2009.

Foucault, M. (2000) *Power*. New York: The New Press.

Gamson, W.A. and A. Modigliani (1989) Media Discourse and Public Opinion on Nuclear Power. A Constructionist Approach. *The American Journal of Sociology* 95(1), 1–37.

Hajer, M. (1995) *The Politics of Environmental Discourse: Ecological Modernization and the Policy Process*. Oxford: Oxford University Press.

Hajer, M. (1996) Ecological Modernisation as Cultural Politics. In S. Lash, B. Szerszynski and B. Wynne (eds), *Risk, Environment and Modernity: Towards a New Ecology*. London, Newbury Park and New Delhi: Sage Publications, pp. 246–68.

Heinonen, J., O.J. Heinonen, J. Manninen and J.K. Miettinen (1978) *Ydinenergia ja elämisen laatu* (Nuclear Power and the Quality of Life). Helsinki: Kustannusosakeyhtiö Tammi.

Hylkilä, I. (2003) *Suomen viidennen ydinvoimalan lobbaus kansanedustajan näkökulmasta* (The Lobbying of Finland's Fifth Nuclear Power Plant from Parliamentarian's Viewpoint). Unpublished Masters thesis in Communication. University of Jyväskylä, Jyväskylä.

Ilmonen, K. and K. Jokinen (2002) *Luottamus modernissa maailmassa* (Trust in the Modern World). Jyväskylä: SoPhi.

Järvelä, M. and M. Wilenius (1996) *Ilmastoriski ja ympäristöpolitiikka. Suomalaiset ympäristövaikuttajat ja nykyajan ympäristöongelmat* (Climate Risk and Environmental Policy. Finnish environmental opinion leaders and today's environmental problems). Tampere: Gaudeamus.

Kantola, I. (2004) Ydinvoimakeskustelun sähköinen kielipeli (Electric Language Game of the Nuclear Discussion). In M. Kojo (ed.), *Ydinvoima, valta ja vastarinta* (Nuclear Power, Power and Resistance). Helsinki: Like, pp. 107–25.

Kamppinen, M., P. Raivola, P. Jokinen and H. Karlsson (1995) *Riskit yhteiskunnassa. Maallikot ja asiantuntijat päätösten tekijöinä* (Risks in Society. The lay people and experts as decision makers). Tampere: Gaudeamus.

Kiljunen, P. (2007) *Energy Attitudes 2006*, http://www.energia.fi/en/publications/finnish%20energy%20attitudes%202006%2c%20research%20report%2026%20june%202007.html – accessed 23 January 2009.

Kojo, M. (2004) Yhteenveto (Summary). In M. Kojo (ed.), *Ydinvoima, valta ja vastarinta* (Nuclear Power, Power and Resistance). Helsinki: Like, pp. 231–58.

Koskinen, K. (1995) Ympäristöpolitiikan hallittu rakennemuutos? Ympäristöpolitiikan rintamalinjat eduskuntakeskustelussa 1980-luvulla (The Controlled Structural Change of Environmental Policy? The front lines of environmental policy in the parliamentary discussion in the 1980s). In I. Massa and O. Rahkonen (eds), *Riskiyhteiskunnan talous. Suomen talouden ekologinen modernisaatio* (The Economy of Risk Society: Ecological Modernization of the Finnish Economy). Helsinki: Gaudeamus, pp. 335–54.

Kyllönen, S. (2004) Ydinvoiman ilmastonmuutos (The Climate Change of Nuclear Power). In M. Kojo (ed.), *Ydinvoima, valta ja vastarinta* (Nuclear Power, Power and Resistance). Helsinki: Like, pp. 51–85.

Lammi, H. (2004) Tarinat kovasta ytimestä (The stories concerning the hard nucleus). In M. Kojo (ed.) *Ydinvoima, valta ja vastarinta* (Nuclear Power, Power and Resistance). Helsinki: Like, pp. 11–50.

Lash, S. (1994) Reflexive and its Doubles: Structure, Aesthetics, Community. In U. Beck, A. Giddens and S. Lash (1994) *Reflexive Modernization: Politics, Tradition and Aesthetics in the Modern Social Order*. Cambridge: Polity Press, pp. 110–73.

Lash, S. and B. Wynne (1992) Introduction. In U. Beck, *Risk Society: Towards a New Modernity*. London, Newbury Park and New Delhi: Sage Publications, pp. 1–8.

Litmanen, T. (2004) Suomen ydinvoimaihme (Finland's Nuclear Miracle). In M. Kojo (ed.), *Ydinvoima, valta ja vastarinta* (Nuclear Power, Power and Resistance). Helsinki: Like, pp. 209–30.

Litmanen, T. (2001) *The Struggle Over Risk: The Spatial, Temporal, and Cultural Dimensions of Protest Against Nuclear Technology*. University of Jyväskylä, Jyväskylä Studies in Education, Psychology and Social Research, No. 177.

Malaska, P., I. Kantola and P. Kasanen (eds) (1989) *Riittääkö energia – riittääkö järki?* (Enough Power – Enough Sense?) Helsinki: Gaudeamus.

Merton, R.K., M. Fiske and P.M. Kendall (1956) *The Focused Interview: A Manual of Problems and Procedures*. Glencoe, IL: The Free Press.

Nissinen, J. (2004) Valtavirran vastarinta (The Resistance of Mainstream). In M. Kojo (ed.) *Ydinvoima, valta ja vastarinta* (Nuclear Power, Power and Resistance). Helsinki: Like, pp. 87–105.

Nurro, M. (1995) *Sähköinen konsensus? Suomen viidettä ydinvoimalaa puoltavan argumentaation analyysi* (Electric Consensus? Analysis of the argumentation in favour of Finland's fifth nuclear power plant). Unpublished Masters thesis of Sociology. University of Tampere, Tampere.

Rosenberg, T. (2004) Kun Loviisa atomisuukon sai (When Loviisa Got an Atomic Kiss). In M. Kojo (ed.), *Ydinvoima, valta ja vastarinta* (Nuclear Power, Power and Resistance). Helsinki: Like, pp. 159–78.

Ruostetsaari, I. (1987) *Energiapolitiikka ja vaikuttajat. Energiapolitiikan määräytymiseen vaikuttaneista tahoista Suomessa* (Energy Policy and Opinion Leaders. About the Influential Actors of Energy Policy Formation in Finland). Ministry of Trade and Industry, Energy Department, Series B; 67, Helsinki.

Sjölander, A. (2004) *Kärnproblem. Opinionsbildning i kärnavfallsdiskursen i Malå* (Core Problem. Opinion Building in the Nuclear Discourse in Malå), Medier och kommunikation nr 7, University of Umeå, http://urn.kb.se/resolve?urn=urn:nbn:se:umu:diva-283 – accessed on 23 January 2009.

Spaargaren, G. (1997) *The Ecological Modernization of Production and Consumption: Essays in Environmental Sociology*. Thesis Landbouw Universiteit Wageningen.

Suhonen, P. (1994) *Mediat, me ja ympäristö* (Media, Us and the Environment). Helsinki: Hanki ja jää.

Suotula, A. (2000) *Energisesti ydinvoimasta. Analyysi Suomen eduskunnan energiapoliittisesta keskustelusta 1990-luvulla* (Energetically about Nuclear Power. An Analysis about the Energy Policy Discussion in the Parliament of Finland in the 1990s). Unpublished Masters thesis in Political Science. University of Tampere, Tampere.

Vahtera, A. (1986) *Ydinenergian uhkapeli* (The Gamble of Nuclear Power). Helsinki: WSOY.

Vehmas, J. (1995) Suomen energiatalouden ekologinen modernisaatio (The Ecological Modernization of Finland's Energy Economy). In I. Massa and O. Rahkonen (eds), *Riskiyhteiskunnan talous. Suomen talouden ekologinen modernisaation* (The Economy of Risk Society: Ecological Modernization of the Finnish Economy). Helsinki: Gaudeamus, pp. 161–88.

Tirkkonen, J. (2000) *Ilmastopolitiikka ja ekologinen modernisaatio. Diskursiivinen tarkastelu suomalaisesta ilmastopolitiikasta ja sen yhteydestä metsäsektorin muutokseen* (Climate Policy and Ecological Modernization: Discursive Study on Finnish Climate Policy and its Connection to the Change of the Forest Sector), University of Tampere, Acta Universitatis Tamperensis 781, Tampere.

Interviews

(Member of the Parliament of Finland/Parliamentary group/Date of interview)
Hassi, Satu Ms/Green League/19 February 2003
Karttunen, Marjukka Ms/National Coalition Party/17 February 2003
Keränen, Niilo Mr/Centre Party/12 February 2003
Kiljunen, Kimmo Mr/Social Democratic Party/12 February 2003
Korhonen, Martti Mr/Left Alliance/16 April 2003
Kärkkäinen, Kari Mr/Christian Democrats/13 February 2003
Kääriäinen, Seppo Mr/Centre Party/11 February 2003
Lax, Henrik Mr/Swedish People's Party/29 April 2003
Paasio, Heli Ms/Social Democratic Party/2 April 2003
Tiuri, Martti Mr/National Coalition Party/9 April 2003
Uotila, Kari Mr/Left Alliance/21 March 2003
Vahasalo, Raija Ms/National Coalition Party/8 April 2003

5

Nuclear Energy Policy Processes in Finland in a Comparative Perspective: Complex Mechanisms of a Strong Administrative State

Erika Säynässalo

Introduction

Since the 1970s objections to the construction of additional nuclear power plants has been one of the main targets of the environmental movements and parties throughout the western world. The anti-nuclear movements and parties have had varying degrees of success in different countries. Anti-nuclear politics has turned out to be successful in Germany where the Red–Green coalition managed to reach an agreement over the phasing out of the existing nuclear energy reactors and also managed to stop the reprocessing of German nuclear fuel in Britain and France. Anti-nuclear politics has also turned out to be quite successful in Sweden; in 1980 the government decided – following the outcome of a referendum – on the gradual phasing out of nuclear power by the year 2010.[1] Finland and France, however, are countries in which anti-nuclear politics has had considerably less success than in Sweden or Germany. During the 1970s the French government managed to resist anti-nuclear protests and to defend its nuclear programme. In Finland in 2002 the majority of the members of Parliament voted in favour of the construction of a fifth nuclear power plant to complement the four existing ones.

The idea for writing this chapter came from the need to understand and explain the particularities and paradoxes of Finnish nuclear energy policy making from a comparative perspective. The logic of the Finnish nuclear energy policy is one of those cases that challenge the credibility of the well-known distinction drawn between strong and weak states.[2] This distinction has turned out to be fertile, especially when trying to capture and explain the differences between the logic of nuclear energy policy in France and the USA. France has typically been regarded as a strong state in its nuclear policy due to its centralized government structures. It has been argued that because of these centralized structures, government agencies and officials have been largely insulated from public pressure. By contrast, the USA has been seen as a weak state because of its fragmented institutional structures that have given the anti-nuclear movement and groups significant levels of access to the state structures (see Katzenstein 1978; Krasner 1984; Skocpol 1979; Cerny 1990).

This distinction is problematic when it is applied to the Finnish case because the pro-nuclear route chosen by the Finnish government cannot be explained simply by referring to the centralized structures of the government. This kind of explanation cannot be given because a specific law, the Nuclear Energy Act (991/1987), was enacted in Finland to authorize Parliament to make the final decision concerning issues related to the further building of nuclear energy plants. However, the position and powers of Parliament have turned out to be anything but unambiguous. As Ari Lampinen indicates in his contribution to this volume, the Ministry of Trade and Industry (MTI) actually had important and influential instruments at its disposal during the decision-making process in 2002, which it used in order to direct the course of this process.

The theoretical point of departure for this chapter comes from the awareness that the logic of nuclear policy making in Finland cannot be made intelligible without placing Finland within the Continental European state tradition. This tradition has resulted in a particular kind of debate concerning the question of the appropriate relationship between the state and democracy (Dyson 1980).[3] To highlight the particularities of the Finnish case, I will present two different answers to this question with the purpose of indicating that these two answers have constituted the basis for the emergence of two types of strong states in the postwar era within the context of this tradition. These two types will be labelled a strong administrative state and a strong party state, respectively. I will analyse France and Finland as examples of strong administrative states and Sweden[4] as an example of a strong party state.

A strong administrative state is a type of unitary state in which an emphasis is placed on the executive branch as a public institution that has been detached from the parliament and which has a source of authority outside parliament (compare Dyson 1980). For this reason, as will be argued below, a strong administrative state creates a foundation for a strong executive branch that is capable of acting independently of the parliament, and on the other hand for a phenomenon that is conceptualized in this chapter as an administrated democracy. This is a form of democracy in which executive power is entrusted to a parliament and possibly also to local institutions in specific issues. The appropriateness of administrative democracy resides in the possibility it provides to politico-administrative elites to view contested and politically important issues as ones that can be settled in the executive arena, rather than in the electoral arena.

It will be argued that the normative structure of strong administrative states has provided effective tools for protecting the continuity of the pro-nuclear route. I therefore put forward the argument that the alliance between the state and the nuclear industry[5] has turned out to be powerful in strong administrative states. I will define the nuclear alliances in strong administrative states as legal-administrative alliances by character.

A strong party state, as will be argued, refers to a state in which the emphasis is not on the legal autonomy of executive power but rather on the capacity of a reformist government party or party coalition to direct public policy making. The interventionist politics of the Swedish Social Democratic Party in the nuclear energy sector between the 1950s and the 1970s can be referred to as an example of the dynamics of a strong party state. As will be argued, the interventionist politics of the Social Democrats constructed the basis for the emergence of a political alliance between the state and the nuclear industry. However, while the legitimacy of this alliance was based on parliamentary party-based democracy, it became possible for the opposition parties to make the nuclear energy issue an electoral issue and to challenge the politics of the Social Democrats.

The aim of this chapter is to indicate that there are more reasons to characterize Finland as an administrative rather than as a party state, especially in the field of (nuclear) energy policy. More specifically, although one cannot argue that the Finnish constitutional-political system has equipped the executive branch of the government with strong legal and political authority in this field as is the case in France, the Constitution of Finland provides normative and practical tools to centralize the decision making of contested and important issues in the executive arena. I will focus my attention on the essence of the decision-making procedure laid down by the Nuclear Energy Act as an administrative procedure. It will be analysed as an example of a source for administrated democracy and as a source for the legal-administrative alliance between the state and the nuclear industry.

The chapter is organized as follows. I will first place the distinction drawn between two types of strong states on the continuum of comparative research on nuclear energy policy making. In this first, theoretical section I will provide a more precise definition of the concept of a strong administrative and party state as well as of administrated democracy and a party-based parliamentary democracy. These conceptual distinctions will be further justified in the following sections of this chapter in which attention will be focused upon the nature of the processes that led to the emergence of alliances between the state and the nuclear industry in three Continental European states, Sweden, France and Finland. The main part of this chapter will be devoted to revealing the logic of the nuclear alliance formation in Finland from the 1950s to the early 2000s.

Comparing the nuclear energy policy processes of three continental European states: Sweden, France and Finland

Conceptual framework for the chapter

In the 1980s, sociologists and political scientists rediscovered the state as a topic of analysis and 'explanatory variable' (Krasner 1984; Evans et al.

1985; Almond 1988; Rockman 1990). In other words, national state structures and their different capacities for action were seen as important 'factors' in the explanations offered for the variance of national strategies in issues related to energy and technology policy or industrial policy (Katzenstein 1978; Campbell 1988).

The theoretical discussion and empirical analysis that revolved around the concept of the state became well-known, especially in respect of the distinction drawn between strong and weak states (Katzenstein 1978; Krasner 1984; Skocpol 1979). This distinction has met with intense criticism among comparative social scientists resulting from difficulties to explain variations in national policies in terms of this distinction (see, for example, Joppke 1992).Yet it is argued here that the problems inherent in these categories of state derive mainly from the tendency of social scientists to regard them as simple categories.

The alternative offered in this chapter is to reveal the Continental European state tradition as the source of a strong state tradition (Dyson 1980). My attempt is to highlight the connection between the historical and theoretical traditions of the state and the present-day political conduct and thereby to distance myself from the tendency of the practitioners of comparative political science to approach political institutions and other structural elements as a catalogue of variables. I will also distance myself from the studies in which scientists ignore the legal context of politics and emphasize informal power or the discretion of the elites as critical 'explanative factors'[6] (for example, Jasper 1990).

By focusing on the Continental European state tradition, I analyse a model that has common features but which is not, however, uniform in terms of its normative structure. It has been common to this tradition to regard the state as a source of collective identity but also as a legal institution with a responsibility for regulating matters of public concern (Dyson 1980, 51). However, there have been differences between the Continental European and Scandinavian states in terms of the ways in which they organize the relationship between bureaucratic and representative power. Dyson (1980) illustrates these differences by way of drawing a distinction between different types of dual polities.

The first type refers to the 'bureaucratic dominance' form of dual policy which has manifested itself especially in the French Fifth Republic of 1958. The issue concerns a dual polity that has contributed to insulating the executive power from the Assembly by way of entrusting the executive branch with greater rule-making authority,[7] by direct election of the president and greater recruitment of bureaucrats to ministerial offices and 'cabinet' positions (Dyson 1980, 64). In this type of a dual polity, politics is viewed as being influenced by particular interests of parties and politicians whereas the executive branch of the government is seen as representing the ability to determine impartially the public interest (Dyson 1980, 64–6).

The other form of a dual policy refers to the 'parliamentary/party dominance of dual polity'. The most well-known contemporary embodiment of this type of polity is (former West) Germany which has often been referred to as an example of a party state. The focus of references has thereby been a state with a strong legacy of distrust of bureaucratic initiatives. It has been parallel to this distrust that efforts have been made to integrate the state and democracy in terms of parliamentary/party dominance.[8]

It should be pointed out that Sweden has not usually been categorized as a party state by political and legal scientists. For example, Dyson categorizes Sweden as belonging to the accommodating polities, characterizing this type of polity as one that has been open to politics, and in which there has been a pragmatic style, a consensual view of social relationships and a recognition of the co-equal and distinctive character of bureaucracy rather than an emphasis on its superiority over other institutions (Dyson 1980, 69).

However, Sweden can be seen as a strong party state if we highlight the important aspect that in a party state the ruling party influences in a critical manner those ideas and norms that govern policy processes. From this perspective, the fact that the Social Democratic Party enjoyed uninterrupted control of the government from 1932 until the elections in 1976 is of critical importance. This provided a foundation for the connection that emerged between the reformist Social Democratic governance, the model of a strong state and the ideas and norms of policy making, such as deliberation, rationalism and openness (see, for example, Ruin 1982).

The strong administrative state, conversely, refers on the one hand to a state that creates a foundation for the centralization of power in the executive branch, and on the other to a phenomenon that is conceptualized in this chapter as administrated democracy. The former phenomenon can be explained by the separation of powers into executive and legislative powers which contributes to an insulation of the executive power from the parliament in a unitary state with centralized government. By administrated democracy I mean a form of democracy in which the authority of the executive is – at least seemingly – narrowed down within the context of specific issues by way of entrusting executive power to a parliament and by way of increasing the number of actors that have access to the executive authority in general. As will be indicated, administrated democracy cannot in any straightforward manner be identified with parliamentary democracy. The question is rather about the form and arena of constitutional politics, because administrated democracy derives its legitimating authority from the distinct legal position of the executive branch and of executive power. Dyson's characterization (1980, 279) contributes to illustrating this point of view. He argues that leaving the question of the appropriate balance between the representative and the state institutions open (as in France) has meant that the fundamental constitutional argument has remained at the centre of politics.

The distinction drawn between a party democracy (in a strong party state) and an administrated democracy (in a strong administrative state) illustrates the issue of analysing the capacity of the governments of strong states to protect the continuity of the chosen line of public policy against the demands and pressures arising from new societal groups. In order for me to illuminate this argument, I focus on the differences between the position and the powers of government in two types of strong states.

In a strong party state government parties hold a visible and critical position in conflict resolution processes which is manifested in the fact that the government's legitimacy depends on the capacity of government party or party coalition to persuade the electorate to support its policies. The heavy reliance on the capacity of the government party or party coalition to gather support for its policy tends to make this policy vulnerable to changes in public opinion and to demands of new societal groups (compare Ruin 1982; Lindvall and Rothstein 2006).

In a strong administrative state a political executive constitutes only one powerful force inside the executive branch, existing alongside the institution of the president and the administration (compare Kesselman 1992, 195). For this reason the legitimacy of a government in a strong administrative state is not critically dependent on the ideological or practical consistency of the activities of government parties. It is often more important for the legitimacy of the government in a strong administrative state that executive power can – at least seemingly – be raised above party politics. For this reason those executive institutions (that is, the president and the administrative organizations) that are perceived as representing the interest of the whole nation may have more critical conflict resolution capacity than government parties. Consequently, political parties are expected to operate under the terms set by the existing bureaucratic and political structures in seeking to advance their interests. It can be concluded on this basis that politico-administrative elites of strong administrative states are more capable than those of strong party states to resist political pressure on the chosen line of public policy.

In the following section I will reveal the different logic underpinning the formation of alliances between the state and the nuclear industry in a strong administrative state and a strong party state during the postwar era. I will also reveal the differences in the logic of the states in facing and dealing with anti-nuclear critique focused on the very acceptability of these alliances. I will analyse Sweden and France as examples of states that provided a foundation for very different nuclear alliances due to the differences in their state traditions and structures. In the final section I will indicate that nuclear energy policy making in Finland can also be seen as an example of a nuclear alliance formation in a strong administrative state.

The discussion on the nuclear energy policies of France and Sweden will be brief because the main focus of the chapter is on Finnish nuclear energy policy.

Strong states and the logic of nuclear policy making: the classic cases of Sweden and France

Sweden: an example of a strong party state

It is impossible to capture and analyse the dynamics of nuclear policy making in Sweden without outlining the dominant position occupied by the Social Democratic party as the governing party in the historical development of the Swedish political system during the nineteenth century. As stated earlier, the Social Democratic Party controlled the government from the 1930s until the elections in 1976. These elections were largely dominated by the nuclear energy issue. This issue contributed to the non-socialist opposition challenging the governing Social Democrats with the result that the non-socialist government was conceived.

The Swedish case reveals the dynamics of a strong party state in many respects. Not only was the Swedish political system dominated by the Social Democratic rule for over 40 years, but the Social Democratic Party was also a reformist party that harnessed the apparatus of a strong state to function as a tool for carrying out the Social Democratic policy of planning and welfare reform in the post-war era (see, for example, Lindvall and Rothstein 2006). The development of nuclear technology suited the Social Democratic model of a strong state and the ideological purpose behind it, which was to advance economic growth and technological change.

The active role of the Social Democratic government in the development of the Swedish nuclear programme took multiple forms[9] and led to the emergence of the so-called 'Swedish line', which was an ambitious plan for the future development of nuclear technology based on heavy water and natural uranium. The active role of the Swedish Social Democratic government – in both the political and the financial sense – in the development of 'the Swedish line' was largely based on its capacity to persuade (both private and public) power companies and industrial enterprises to co-operate with the state. The influence of the Social Democrats in the development of this co-operation was so strong that we can speak of the emergence of a political alliance between the state and the nuclear industry. This alliance culminated in the enactment of the Atomic Energy Act in 1956. The law proposed the Swedish nuclear reactor line in addition to setting up the regulatory apparatus for the licensing and control of nuclear reactors (see Jasper 1990).

However, the Social Democratic government did not turn out to be all-powerful in the nuclear sector. At the end of the 1950s it could clearly be seen that the Social Democratic government had not managed to defend the 'Swedish line' against the interests of the Swedish industry, with the result that the government withdrew from the 'Swedish line'.[10] The political line that was withdrawn was the promotion of the politics of national independence. Yet the Social Democratic government continued pursuing a policy of active state intervention in the nuclear industry.[11]

In the early 1970s the Social Democratic politics of a strong state became the focus of intense opposition critique.[12] The Centre Party managed to make the nuclear issue a partisan issue that symbolized the differences between the two parties' attitudes towards the environment – a strategy that reduced the plausibility of the Social Democrats' defence. It is logical to conclude that the rise of the Centre Party as a channel of anti-nuclear opposition offers a substantial explanation of the weakening legitimacy of the Social Democrats' nuclear programme from the mid-1970s onwards. In addition, the nuclear issue divided the Social Democrats within the party (compare Jasper 1990, 139).

However, we can pose the question of what explains the success of the Centre Party's opposition politics. In my view, it is not enough to refer to the strategies utilized by the Centre Party. It appears to me that the success of the opposition politics of the Centre Party can be traced back to the process through which the Social Democrats developed into an influential government party that gave 'public face' to the strong interventionist state in the nuclear energy sector. It was as a result of this interventionist line that the alliance between the state and the nuclear industry became openly political and therefore open to political critique.

It is somewhat paradoxical that the Social Democrats also themselves generated the foundation for the collapse of their pro-nuclear politics – as well as for their hegemonic position in a broader sense – along with carrying out electoral and constitutional reforms in the early 1970s. These reforms eliminated the Upper House of Parliament[13] and made the electoral system nearly fully proportional,[14] resulting a dramatic weakening of the capability of the Social Democrats to gain and sustain a majority in Parliament.

In my view then the dynamics of nuclear policy making in Sweden contribute to illustrating the operation of a strong party state. First of all, the state structures allowed the emergence of a reformist government party that gave 'public face' to the interventionist state. Secondly, after the introduction of the new Constitution in 1974 the Social Democratic governance became susceptible to changes in electorates and thereby to strong opposition politics that was directed against this governance. Thirdly, the collapse of the Social Democratic pro-nuclear politics after the 1976 election caused the Social Democrats to re-evaluate the line of the party and finally to their party's orientation becoming anti-nuclear.[15] This ideological shift indicates that the need to protect the ideological integrity is of importance for the legitimacy of parties in a strong party state. The public image of political parties cannot bear up under strong internal opposition and division.

From the early 1980s onwards political parties in Sweden largely committed themselves to a programme to phase out nuclear energy, although the capacity of government parties to impose anti-nuclear reform policies in accordance with the chosen schedule turned out to be limited. In February 2009 the bourgeois government decided to abandon the phase-out policy and to allow the construction of nuclear reactors in Sweden. The new political line is related to the requirements set by climate change for national energy policies. It will be seen in the future whether the further construction of nuclear power plants might become a reality in Sweden.

It would be helpful to compare the logic of the nuclear energy policy-making process in Sweden to that in Germany. In both of these states the nuclear energy policies became intertwined with the reformist politics of the Social Democrats and in both of these states the Social Democrats were forced to change their nuclear energy positions in the face of strengthening anti-nuclear protests. Both of these countries have also committed themselves to the phasing-out programmes which have turned out to be difficult to implement in accordance with the chosen schedule. However, there is no scope for this kind of comparative analysis in this chapter.

France: an example of a strong administrative state

The foundation for the alliance between the state and the nuclear industry in France is the constitution of the Fifth Republic of 1958. This constitution, as stated above, has laid down the basis for the 'bureaucratic dominance' model of dual polity. I have conceptualized the French state as a strong administrative state.

The constitutional reform that gave birth to the Fifth Republic was influenced by the need to rethink the role of the state in directing the economy in general, and in directing industrial, scientific, and technological development in particular. Large numbers of state institutions were established or reformed prior to the constitutional reform. The Planning Commission was the most important institution, reflecting the new activist approach which sought to unite dynamic, private business groups and state administrators in the common task of modernizing French capitalism. The profusion of new and reformed institutions led to the emergence of a growing class of state experts that became aware of their role in public service with the alleged disinterest of the nation-state. The emergence of a new class of state experts culminated in a restructured state and especially in the position that was given to nuclear energy as one of the fields of industry that became nationalized. The result was a strong connection that emerged between the development of the nuclear sector and a powerful ideology that equated the production of nuclear energy with the protection of national independence (see, for example, Hecht 1998; Nelkin and Pollak 1980; Jasper 1990).

The argument that the alliance between the state and the nuclear industry in France was legal-administrative from its inception is not an exaggeration, because the nuclear industry gained a special position as a state industry with the result that it gained more authority than the political executive over the direction of nuclear policy (see Jasper 1990). Yet it should be stressed that the strong commitment to the development of nuclear energy was also in the interests of the conservative governments that were in power from the end of the 1950s to the early 1980s.

In particular, the state-owned utility EDF[16] managed to gain an influential position within the state. It gained support from the PEON commission[17] for its interest in expanding the nuclear programme in the early 1970s. Yet it was the massive Messner Plan, approved by the government in 1974, that unleashed the technological enthusiasm in France. This plan called for EDF to begin the construction of 13 new reactors in 1974 and 1975 with the purpose of committing itself fully to nuclear energy.[18]

It is obvious that the Messner Plan was not a result of the political goals of Messner's government. The purpose behind this plan was rather to receive an affirmation for a plan that already existed, independently of the goals of Messner's government. However, political leaders gained authority to make some strategic choices. Prime Minister Messner together with President Pompidou took the side of technologists in the conflict between EDF's ambitious plan and the Finance Ministry's cost–benefit perspective.[19] In other words, EDF needed political support in order to be able to damp down the critique addressed by the Finance Ministry against the ambitious plan. In addition, it can be concluded that the Messner Plan offered the government the possibility of presenting the plan as a French answer to the oil crisis; an answer that offered the only path to energy independence (see Jasper 1990, 156).

Although there were also other European governments – for example, the governments of Sweden and West Germany – that had provided an expansion of nuclear energy as a response to the oil crisis, it was peculiar to the French case that the commitment to nuclear energy was massive. One explanation for this is that there were no reasons for the conservative government to resist the announcement of the plan because 'the planning, enthusiasm and technical infrastructure were already in place, just waiting to be put in motion' (Jasper 1990, 156). It can be regarded as a more crucial aspect that there were no signs of the emergence of a strong opposition party that could have channelled the anti-nuclear demands into the state. Over the course of the 1970s it became evident that all of the main parties supported nuclear energy. For example, the main opposition party, the Socialists, supported the expansion of nuclear energy because it was in line with their pro-growth ideology. Hence, the Socialists did not adopt a clear anti-nuclear position resulting in their not trying to call into question the acceptability of nuclear energy as such. Instead, they focused their critique on the size of the Messner Plan, arguing that it was favouring the interest of big business. The Socialists also argued that Parliament was entirely ignored when this plan was conceived (Jasper 1990).

It can be concluded that the Socialists applied a kind of dual strategy – criticizing the Messner Plan without criticizing the extension of nuclear energy as such. That is, the Socialists connected the strategy of the opposition party to the strategy of a potential forthcoming government party that was oriented to adapt itself to the prevailing bureaucratic and industrial structures. Referring to this dual strategy is of particular importance because the Socialists gained a position as a mediator between the state and the anti-nuclear opposition at the end of the 1970s. This position of a mediator was apparent when it became clear to the anti-nuclear groups that gaining a Socialist government was the only way for them to try to advance their anti-nuclear line (see Jasper 1990).

The behaviour of the Socialists after they came to power in 1981 illustrates the assumption set forward above that parties in strong administrative states tend to adapt themselves to the existing bureaucratic structures. The Socialists made only minor curtailments in the Messner Plan, although there were grounds for more radical curtailments because the forecasts of electricity demand growth showed that the level of demand was shrinking (Jasper 1990, 247). However, the Socialist government strengthened the prerequisites for political participation and began to produce reports and to establish committees designed to gather information and opinions. The government suggested that decisions concerning particular sites would go through three tiers of approval if necessary: first the municipal councils, then the regional ones and finally the National Assembly itself. In addition, there would be a genuine debate on energy policy in the National Assembly. Over the course of the following years the Socialists also brought representatives of consumers and small builders into the energy planning commission (see Jasper 1990, 247).

Jasper (1990, 247) concludes that these formal changes had little effect on the implementation of the nuclear programme. It should be stressed, however, that the measures taken by the Socialist government contributed to an indication that the capacity of the French government to advocate an ambitious nuclear programme was not based mainly on its unambiguous ability to impose decisions notwithstanding opposition from groups nor from repressive measures that were utilized to put down anti-nuclear opposition.[20] While it was clear that the administrative-industrial elite required the support of the political elite for its most ambitious plans, these plans needed to be legitimated to the wider public.

The concept of administrated democracy helps to explain the measures taken by the Socialist government to strengthen democracy. Although administrative reforms were introduced in the nuclear sector with the aim of reducing the administrative power of central government and strengthening the power of Parliament as well as local and regional governments, the reforms were not intended to adjust the existing balance of power in any fundamental manner. Conversely, there are reasons to argue that the administrative reforms were a strategy for the Socialists to turn important political and contested issues into largely constitutional issues. This argument gains empirical support from the fact that after they assumed power in 1981 the Socialist Party leaders suppressed the anti-nuclear opposition within the party.[21] It has been concluded that the capacity of the Socialist Party leaders to marginalize anti-nuclear groups in the party was a label that could stick. As a result, the French anti-nuclear movement was finished for a long period (see Jasper 1990, 249).

To understand the strategies of the French Socialists in quelling the anti-nuclear opposition inside the party, we also need to focus our attention on the structure of the French Socialist Party. Unlike its counterpart in Sweden, this was not a party characterized by a mass organization and grassroots activists. Respectively, the Socialist Party in France did not need to be concerned with ideological cohesion as a prerequisite for its credibility in the eyes of voters as was the case in Sweden. Parallel to this aspect was the fact that the legitimacy of the alliance between the state and the nuclear industry was not in any fundamental sense dependent on the consistency of the Socialist Party's politics in France.

Therefore it can be concluded that the success of the French nuclear programme can be explained by two aspects. It was based on the legal-administrative alliance between the state and the nuclear industry, resulting in the state organizations, such as EDF, being able to count on the support of the main part of the centralized administration. Secondly, the administrative-industrial elite was able to form a successful alliances with the presidents and the governing political parties.

The dynamics of nuclear energy policies in Finland: a perspective on the disguised operation of a strong administrative state

Why Finland can be seen as a strong administrative state

At first sight there is nothing very surprising in characterizing Finland as a strong administrative state. Finland, just like both Sweden and France, is a centralized state with a strongly professionalized civil service. In addition, the Finnish state – as strong states in general – has played an important role in creating a foundation for the industrialization of society. However, there are less clear or unambiguous reasons to argue that the Finnish Constitution contributes to making Finland an administrative rather than a party state due to setting down the basis for the separation of the executive and legislative powers.

It has become rather commonplace in Finland to see the separation drawn between the legislative and executive powers as a constitutional principle that contributes to the protection of democracy. This conclusion has been drawn on the basis of the argument that the centralization of power in the hands of one single body – Parliament – would damage democracy. The fact that the separation of powers principle is difficult to reconcile with the principle of parliamentarianism – the other central principle of the Finnish

Constitution – is a fact that has not provoked deep-grounded theoretical discussion among legal and political scientists. This kind of a conclusion can be put forward although the Constitution of 2000[22] was enacted mainly for the purpose of strengthening the parliamentary features of the Finnish constitutional and political life (see, for example, Jyränki 2003; Tuori 2005).

Two reasons can be referred to when an answer is sought to the question of why the separation of powers principle has not provoked deep-grounded discussion among Finnish political and legal scientists as a principle that narrows the scope of parliamentary democracy. Firstly, a strong executive has often been associated with the strong powers of the Finnish president. It is logical from this perspective that the constitutional reform conducted in the 1990s was strongly influenced by the need to reduce the powers of the president. The new constitution of 2000 contributed, for example, to reducing the ways in which the president had earlier interfered in the government formation process as well as in other governmental activities (see Nousiainen 2001; Paloheimo 2003). In addition, the new constitution brought the legislative power under the authority of Parliament while the old constitution had vested the president with the independent power to issue statutes.

Secondly, the old Constitution of Finland, which was introduced in 1919, laid down strong rights for minorities in Parliament with the purpose of preventing the emergence of an overly powerful executive branch. According to the old Constitution Act, any law enacted in Parliament could be postponed if one-third of all the members voted for it.[23] The paradox is, however, that this strong power instrument available to the minorities of the Parliament was directed principally against the political executive. In other words, this power instrument actually prevented the emergence of strong working majorities in the executive branch. Short-lived coalition governments were commonplace in Finnish politics until the end of the 1960s. The unstable governments in turn required other state institutions – such as a president and a bureaucracy – to fill the breach. It could be argued then that equipping the minorities of Parliament with a strong instrument of power offers one explanation to the question of why Finland cannot be viewed as a strong party state. This brief remark can be illustrated by referring to the original purpose of these minority rights which had been to prevent radical socialist reforms and revolution being implemented by a simple majority.[24]

However, as early as the 1980s – that is, before the constitutional reforms – the government's stability was strengthened considerably within the framework of the three-party system (Conservatives–Agrarians–Socialists). Consequently, governments have shown a far stronger commitment than before to their programmes and have also exhibited less tolerance to interference from outside actors. The stability of the governments was further strengthened in the early 1990s when the rights of minorities in Parliament to postpone legislation were annulled by the amendment to the Constitution. This critical step towards the so-called normal parliamentarianism in the early 1990s was completed in 2000 by the means of the Constitution that reduced the powers of the president (see Nousiainen 2001; Paloheimo 2003).

It is concluded here that the formal constitutional changes have not led Finland to take a step towards being a parliamentary party state. It is more reasonable to speak in terms of a step that has been taken towards being an administrative state led by strong ministries. This argument in its entirety cannot be justified within the boundaries of this chapter, but it receives support from three interrelated elements of the Finnish constitutional-political system.

Firstly, the decline of ideological parties has been evident in Finland, manifesting itself especially in the lessening significance of traditional ideological lines in the formation of governments (see Paloheimo 2003). For example, the governments that were formed in the mid-1990s onwards were coalitions that cut radically across the right–left division.[25] Secondly, while the broad ideological issues have vanished, the political agenda have become dominated by issues that are treated as technical or administrative rather than ideological. This in turn has strengthened the powers of the administrative and technical elites who have been seen as supplying the expertise needed for the initiation, formulation and implementation of policies.

The third – and perhaps the most important – reason for placing Finland in the category of an administrative state rises from the particular kind of position of Parliament as an institution that can be entrusted with the authority to make an administrative decision in particular cases. This position has generated a normative foundation for a phenomenon that is conceptualized in this chapter as administrated democracy.

The role of Parliament in the decision making in relation to the budget contributes to illustrating its role as the exerciser of administrative authority. The procedure for determining the budget is the only one recognized by the Finnish Constitution that authorizes Parliament to make a decision that is administrative by its legal character. The widely shared view among Finnish administrative scientists has been that while budget power has been 'merely' a legal instrument at the disposal of Parliament, the actual political influence arising from the budget procedure has been concentrated in the Ministry of Finance (see Tiihonen and Tiihonen 1990; Tiihonen 1990; Harrinvirta 1997).

It can be argued that the budget procedure represents a mechanism through which a realization can be given to two controversial constitutional goals: Parliament's control over administrative power and legal-bureaucratic control over the political power of Parliament. The question is about Parliament's control over administrative power because it is Parliament that makes the formal decision concerning the state budget. Yet the question is also about legal-bureaucratic control over Parliament because the procedure sets tight legal limits on the actual leeway of Parliament. The budget power of the Finnish Parliament is not thereby a 'normal' legislative or parliamentary power by character. Although it authorizes Parliament to make a decision on the budget, it can also be seen as obligating Parliament to make this decision within the given set of legal restraints. The budget procedure represents thereby a form of administrated democracy in which law and administration – not party-based parliamentary process – play a major role.

There are similarities between the budget procedure and the procedure for decision making of nuclear energy laid down by the Nuclear Energy Act of 1987. In both cases Parliament is entrusted with administrative authority. While the budget procedure has contributed to making the Ministry of Finance an influential actor in the field of public economy and finance policy, it can be argued that the Nuclear Energy Act has contributed to a strengthening of the position and powers of the Ministry of Employment and Economy (the former Ministry of Trade and Industry) in the field of (nuclear) energy policy. In the next section I will next move to specify the reasons for this kind of development.

The changing role of the state in the nuclear energy sector of Finland

From the 1950s to the 1970s – from a weak state to a strong administrative state

When the early history of nuclear energy politics in Finland is compared to the corresponding periods in France and Sweden, it becomes evident that the contribution of the state to the development of the nuclear sector was clearly less ambitious and therefore weaker in the case of Finland. There are several reasons that explain this weaker role.

First, industrialization had advanced at a slower pace in Finland than occurred, for example, in Sweden. In addition, Finland had less capital to be invested in the development of nuclear energy than was the case in Sweden. Furthermore, the know-how related to nuclear energy was largely absent in Finland. Second, the Peace Treaty of Paris denied Finland the possibility to acquire and produce nuclear weapons, so Finland was not allowed to have any interest in the military aspect related to nuclear power. Third, the role of the state in electricity production had turned out to be problematic. During the 1950s the state-owned electricity companies established to build hydroelectric power (for example, Imatran Voima Oy, Kemijoki Oy) had conflicted with the interests of private power companies. These conflicts revolved around the question of ownership and control over hydroelectric power. The core of these quarrels concerned the terms under which the forest industry could acquire cheap electricity. Hence, the relationship between the state-owned electricity companies and private power companies was already tense when the debate on nuclear energy started in Finland.

Fourth, it turned out to be difficult for the governments to steer state-owned companies into following the line set by the governments. Following their establishment, the power companies tended to choose their course of action largely independently of the political lines advanced by the government (see Karjalainen 1989). There are two interrelated aspects that explain this fourth issue. On the one hand, the Finnish constitutions have not equipped the government with a strong rule-making authority. Thus, the legal-administrative prerequisites for the Finnish government to take an active role in the development of the economy and industry have been weaker than in France where administrative organizations gained a leading role in the modernization agenda. On the other hand, the Finnish governments lacked a clear political ideology in the post-war era, resulting largely from the prevalence of weak and ideologically divided coalition governments as well as from legal restraints on the government's capacity to decide on economic and industrial issues (Karjalainen 1989). For example the socialist–agrarian coalition governments that were the most frequent type of government during the 1950s were largely incapable of taking a leading role in the economy and energy politics.

The fact that the Finnish state lacked ambitious nuclear strategies in the 1950s can be illustrated by referring to the content of the 1957 Atomic Energy Act. This law reflected an ideology of economic liberalism by way of regarding the use of nuclear energy as a form of entrepreneurship that 'only' needed a special kind of a license in order to be conducted. The permit conditions were clearly looser than those laid down by the Atomic Energy Act of 1956 in Sweden. The Finnish law gave virtually free reign to the industrial companies for the construction of NPPs.

The position and powers of the state in the nuclear sector also remained ambiguous throughout the 1960s. During this decade some committees and subcommittees were established under the formal authority of the MTI to coordinate the process of acquiring and constructing NPPs. Yet it was evident that the authority of these institutions was unclear; an aspect that was parallel to the interests of both the state-run company IVO and the private industry to advance the acquisition of NPPs independently of state involvement. It was not until the end of the 1960s that the industrial actors became aware of the need for stronger state involvement. It became evident at that time that the acquisition of the first NPP was intertwined with complex questions related to both foreign policy and trade (see Sunell 2001, 2004).

The acquisition of the first NPP was arranged by means of presidential intervention.[26] It has been argued that the president's support also made it possible for the private-owned company TVO to acquire NPPs.[27] Despite the presidential interventions into the nuclear energy sector, the visible role of the state in the nuclear sector was weak rather than strong throughout the 1970s. It can be regarded as a paradoxical feature of Finnish nuclear energy policy during the 1970s that coalition governments avoided taking a position concerning the question of the fifth nuclear reactor. It was a paradox because it was clear from the early 1970s onwards that the nuclear energy companies were willing to build the fifth nuclear plant (see Michelsen and Särkikoski

2005) and this development, coupled with the first oil crisis in the mid-1970s, helped to push energy issues up the political agenda, forcing governments to compile comprehensive energy programmes. Contrary to the energy programmes of France, Sweden and West Germany, which responded to the oil crisis by committing themselves to the implementation or extension of their nuclear programmes, the official line of the Finnish government was not to commit openly to nuclear energy. Nuclear energy was just mentioned as one of the possible energy sources to be explored further whereas the government announced that it would take an active role in promoting the utilization of domestic energy sources (see, for example, Ruostetsaari 1986).

The explanation for the noncommittal line of the Finnish governments during this period should be sought from two aspects. Contrary to France, where administrative organizations were entrusted with the authority to carry out the economic and technological modernization of the country and where the nuclear industry occupied a state monopoly, Finland lacked an energy administration equipped with a corresponding institutional status and a clear ideological role. Neither were the fragmented party governments able to openly pave the way to the commitment to the extension of nuclear energy.

However, there are justified reasons to argue that the emergence of a strong administrative state in the nuclear energy sector was detectable as early as the 1970s. Unlike in France, the strong administrative state in Finland did not take the form of an assertive government capable of carrying out the politics of imposition. The particular Finnish mode of an administrative state derived rather from the administrative procedure for decision making concerning nuclear energy due to its role as a framework under which the discussion of the appropriateness of the further construction of nuclear energy took place. The foundation for this procedure was laid down by the old Atomic Energy Act of 1957. According to this law, the application for the construction of single NPPs was to go through a particular kind of a licensing process.

It was evident in the early 1970s that the licensing procedure set down by the old Atomic Energy Act contributed to the Finnish politico-administrative elite treating the political challenges raised by the nuclear energy issue as one that concerned constitutional politics. The question at stake from the early 1970s onwards was who can or who should be entrusted to hold the position of the supreme authorizing body in making decisions concerning the construction of nuclear plants in Finland. The Atomic Energy Act of 1957 put the MTI in a position to exercise this authority while the socialist–agrarian coalition government led by Prime Minister Kalevi Sorsa (Social Democrat) took the position in its programme that the acquisition of NPPs ought to be made dependent on the decision of principle made by the government (see Sunell 2001, 2–3). In other words, it was possible for the politico-administrative decision makers in Finland to control the increasing political sensitivity of nuclear energy by viewing it as a question that concerned the choice of an appropriate decision-making body. The decision to entrust the government to hold the position as the supreme authorizing body was constitutional politics at its most basic level. The political dimension inherent in the nuclear energy issue became associated with the acknowledged need to make the decisions concerning nuclear energy at a high political level.

What I am also arguing is that the possibility to conceive of the nuclear energy issue as one that concerned the proper authorizing body inside the executive branch was one that offered to the politico-administrative elite the ability to view the nuclear energy issue as an administrative one by character. The nature of the nuclear energy issue as an administrative issue largely explains the opportunity the nuclear companies had to advance the plan for the construction of the fifth nuclear plant during the 1970s and early 1980s[28] without there being open political authorization for these plans. It can be seen as having been in the interest of the nuclear industry to gain the acceptance of the government for the plan for the construction of the fifth nuclear reactor by way of following the formal licensing procedure.

It can be concluded that the possibility to label the further construction of nuclear energy as an administrative issue that could be treated by means of the administrative-legal procedure contributed to liberating government parties from answering sensitive questions related to the acceptability of nuclear energy as an energy source. In other words, the nuclear energy issue was placed outside the party-based parliamentary arena. Hence, what is argued here is that while the noncommittal line of the Finnish government in the nuclear energy issue appears at first sight to be a sign of a weak state, a closer examination reveals that this was not necessarily the case. As will be indicated, the question was about a line of constitutional politics that contributed to neutralizing the nuclear energy issue from strong ideological party confrontations such as those that led to the collapse of pro-nuclear politics in Sweden and (West-)Germany from the 1970s onwards. In Finland, the possibility for the further building of nuclear energy remained an option available to the nuclear industry even after the Chernobyl accident of 1986. This was largely the case because no previous government had adopted the role of promoting nuclear energy.

The fact that the nuclear energy issue did not constitute a clear party-political issue in Finland contributes to an explanation of the enactment of the Nuclear Energy Act. In the next section this law will be analysed as an affirmation of the legal-administrative alliance between the state and the nuclear industry in Finland.

The enactment of the Nuclear Energy Act of 1987

The plan to revise the appropriateness of the nuclear energy legislation was raised in the mid-1970s,[29] but the reform did not take a critical step forward until after the Chernobyl accident in 1986. The new Nuclear Energy

Act of 1987 laid down tighter licensing conditions for the construction of new reactors than had been given in the old Atomic Energy Act of 1957. In addition, the new law contributed to an increase in the number of actors who were allowed to take part in the decision-making process concerning the construction of nuclear power plants.

The decision-making procedure laid down by the new law contained the public hearing of local citizens, municipalities and public authorities (see the chapters by Kojo and Lampinen in this book). While the old law did not allow for local inhabitants and authorities to express their opinions and viewpoints to the licensing authorities concerning nuclear sites, the new law marked a clear improvement. Yet, compared to the nuclear legislation of the other western countries, this procedure for organizing public hearings was not very exceptional.[30]

What was exceptional in the new Nuclear Energy Act was the role offered to the Finnish Parliament. This institution was granted the final authority to decide on the acceptability of the Decision-in-Principle (DiP) made by the government. More precisely, the law requires that the government evaluates the licence application from the perspective of the 'overall good of society as a whole' after which it could make a DiP concerning the application. The decision then would need to be ratified by Parliament.

It is easy to confuse the thought of democracy underlying the decision-making procedure laid down by the Nuclear Energy Act with the idea of parliamentary democracy,[31] because the law in question recognizes the importance of Parliament as a decision-making arena. However, the law restricts Parliament's opportunities to influence the content of the DiP. It is possible for the majority of the MPs only to accept or to reject the DiP of the government. In addition, what is of particular importance is that the law in question indirectly prohibits the government from taking the initiative in nuclear energy policy making. It appears rather that the Nuclear Energy Act harks back to the old Atomic Energy Act of 1957 in that it provides the NPP applicant with a significant position in influencing the nuclear policy of the government by way of giving the applicant the possibility of deciding when and if to bring the application for the new reactor onto the political agenda. Hence, the new law also takes the view that the nuclear company has a legal right to apply for the DiP and that the constitutional institutions – the government and Parliament – have to treat the application in accordance with the licensing procedure laid down by law.

This chapter has argued that the concept of administrated democracy is required in order to explain the view of democracy underlying the Nuclear Energy Act. The law grants Parliament administrative authority, reflecting an understanding that this authority does not belong to Parliament, but that it can be entrusted to Parliament in a single case.[32] The idea is somewhat the same as that applied by the French Socialists after they assumed power in 1981. They gave Parliament a more visible role as an arena of nuclear debate and decision making than had been the case under their predecessors, the conservative governments. From the contemporary perspective the Socialists' attempts to strengthen democratic elements in the nuclear energy policy of France appear as attempts to legitimate the commitment to the pro-nuclear route rather than as attempts to provide a real possibility for alternative viewpoints to gain access and influence inside the state structures (compare Jasper 1990).

In Finland, the authority of Parliament in the decision making of nuclear energy was determined by law, meaning that the Parliament was officially entrusted with an administrative task.[33] Yet, when the Government Bill for the Nuclear Energy Act (16/1985) was under discussion in the 1980s, no consensus prevailed among the MPs concerning the influence Parliament would receive under the new law.[34] The Nuclear Energy Act was a compromise, especially in relation to the position and powers of Parliament. The complexity inherent in the constitutional and political position of Parliament became apparent in the early 1990s when the first nuclear energy policy process based on the new law took place.

Explaining nuclear energy decision making in Finland from the 1990s onwards

The first decision-making process concerning the further construction of nuclear energy based on the new Nuclear Energy Act began in 1991 when the joint company Perusvoima Ltd[35] applied for the DiP. It was characteristic of this political process from the very beginning that political parties were divided in respect of the nuclear energy issue. It also indicated that the nuclear company was not prepared for the political complexity that the fragmentation of the political camp brought about. The representatives of the company considered the opportunity for the construction of the fifth nuclear power unit to be favourable because the Minister of Trade and Industry, Ilkka Suominen from the National Coalition Party, clearly supported the further development of nuclear energy. Less attention on behalf of the nuclear company seemed to have been focused on the fact that the Prime Minister, Esko Aho from the Centre Party, was opposed to the construction of the fifth NPP or on the fact that the government had not defined its position on the nuclear issue in its government programme.

The insufficient preparation for the complexity of the political setting on behalf of the applicant company was reflected in the events that occurred in the Autumn of 1992 before the DiP prepared by the government had been submitted to Parliament. At that time the government's account for the energy strategy of Finland was under discussion in Parliament. The reading of this account in Parliament led to voting in which the majority of the MPs (which included members of both the government and opposition parties) took a position that excluded the further building of nuclear power plants from the Finnish energy strategy.[36] This decision of the majority of the MPs to resist the further building of nuclear energy by using channels other than those offered by the official procedure and thereby to pass over this procedure was one that had built into it a possibility that the anti-nuclear opposition in Parliament might have taken the initiative in nuclear energy policy. The initiative in the hands of Parliament would have indicated that the official procedure enacted for the purpose of deciding on the further building of nuclear energy as a separate issue would be brought down.

In early 1993 the chaotic situation led government ministers to adopt an interpretation that the decision about the nuclear reactor will be made on the basis of a free rein granted to individual ministers. The majority of political parties had already earlier made a decision to grant a free hand to MPs in relation to this issue. The possibility available to political parties and politicians to turn the nuclear issue into an 'issue of conscience' liberated government parties from having to take a position on the nuclear issue. It is somewhat a paradox that this possibility contributed to delivering the credibility of the procedure laid down by the Nuclear Energy Act. This kind of conclusion can be made when we take into account that the application for the fifth NPP almost failed due to the apparent inability of political parties to create cohesive voting in the government and Parliament on the nuclear issue.

The first decision-making process based on the Nuclear Energy Act of 1987 in the early 1990s can be seen as being twofold by its consequences. On the one hand, the majority of MPs rejected the plan for a new NPP in 1993 and this decision seemed to indicate the importance of Parliament as a channel for the voice of nuclear opposition to be heard.[37] This decision led to the plan for the further construction of nuclear energy being buried for a decade. On the other hand, however, the process had followed the official procedure laid down by the Nuclear Energy Act. The process thereby also contributed to justifying one of the most fundamental principles of the law according to which the nuclear company has a legal right to apply for the DiP and that the constitutional institutions have to treat the application in accordance with the licensing procedure laid down by the law in question. This principle had come to mean that the government was expected to take a position on the application, even if it was unable to take a politically unified position. The view of the nuclear issue as an 'issue of conscience' was well suited to legitimate administrative legalism built into the official procedure. In this light the events of 1992 reflected the weak legitimacy of party-led parliamentarianism in the Finnish constitutional and political tradition.

That the nuclear company TVO renewed an application for the construction of the fifth NPP in 2000 provides support for the argument advanced above. The previous process did not contribute to a calling into question of the legal-administrative alliance between the nuclear industry and the state based on the Nuclear Energy Act. In addition, the second process contributes to an illustration of what is actually meant by the applicant company having the instruments of policy initiation at its disposal. It was apparent from the beginning of the second process in the early 2000s that this process had been more carefully prepared on behalf of the pro-nuclear camp than had been the case in the first one. Prime Minister Paavo Lipponen of the Social Democratic Party supported the plans of the TVO, and so did the Minister of Trade and Industry, Sinikka Mönkäre. The main political strategy for convincing representatives and the wider public was to view nuclear energy as a means by which Finland could respond to the need to reduce greenhouse gases and so reduce the impact of climate change. It has been argued that the plan for the fifth nuclear power plant was framed as environmentally friendly by TVO and the heavy industries that supported them (see Lampinen, chapter 2 in this volume; Kyllönen 2004).

My intention is not to move into a closer analysis of the political argumentation and strategies that were utilized by the applicant and the wider pro-nuclear camp in their efforts to justify the appropriateness of the additional nuclear energy (see the chapters by Lampinen and Lammi in this book). Political aspects are of importance, but the political strategies available to the pro-nuclear camp can be made understandable only by focusing attention on the particularities of the decision-making setting. In this perspective it is of particular importance to notice that from the beginning the government was liberated from a demand for a consensus, marking that there was no pressure put on it to formulate a unified line in the nuclear energy issue.

It can be concluded that the decision-making rules of the government have become fragmented in the energy sector from 1993 onwards. There has been majoritarian and consensual decision making. The government follows the majoritarian decision-making rule when it decides on the DiP concerning the application for a new NPP. In this case the government does not act as a politically unified actor. It is possible for the government to follow the consensual decision-making rule – which means taking an unified position – when the issue to be decided does not divide the government. For example, when the government introduces a bill or presents its energy strategy to Parliament it acts as a unified political actor, as a 'consensus government'.

It is obvious that when a government's decision-making rules are fragmented in a particular policy area, it cannot be a strong political actor. Although individual ministers can occupy a strong position, this does not make the government strong as a whole. It is paradoxical that the situation in the early 2000s – when the TVO's application for the fifth nuclear power plant was under evaluation – did not differ much from the earlier history of (nuclear) energy policy making, of which weak governments have been characteristic. It was characteristic also of the policy process of the early 2000s that the government was ideologically divided and had no capacity for unified action.

At first sight there appear to be few similarities between the strategies applied by the French Socialist government in the early 1980s and those adopted by the coalition government led by the Social Democrats in Finland in the early 2000s. In France, the Socialist Party leaders quelled the critical opposition inside the party after having come to power whereas in Finland government parties played no open or visible role in ensuring government discipline because parties had earlier given free rein to the MPs in votes related to nuclear energy issue.

Yet, despite these obvious differences between the strategies applied by the government parties in these two countries, their strategies reflected similar problems. In both cases political parties have had hardly no choice other than to adapt themselves to the complex legal-administrative framework. In the French case the legal-administrative framework consisted of a powerful bureaucracy. The nuclear companies (such as EDF) constituted one part of the state machinery and could thereby influence, in a critical manner, the alternatives and options that became available to government parties. These circumstances can be seen as having encouraged the Socialist Party leaders to adopt repressive strategies against nuclear critique inside the party.

In Finland it was the decision-making procedure established by the Nuclear Energy Act that constituted the complex legal-administrative infrastructure within which parties had to operate. As has been argued, this law obligated politicians to decide upon the application for the DiP in accordance with the formal procedure. As was also argued above, the law indirectly gave birth to the fragmentation of the decision-making rules of the government. The structural weakness of the fragmented government has been manifested in that the ministry in charge of the coordination of energy policy (the MTI) has had strong instruments of power at its disposal. Lampinen's chapter in this book reveals how MTI was able to influence those alternatives for energy strategies that were submitted to Parliament for deliberation in the process that took place in the early 2000s.

It has naturally been common to the logic of both French and Finnish nuclear energy policy making that Parliament has become impotent as a channel for the nuclear opposition to gain influence in the line of nuclear policy making. In France the Parliament had no effective constitutional and political instruments with which to try to influence the powerful administrative-industrial infrastructure that revolved around the nuclear industry. In addition, party discipline was used efficiently by the Socialist Party leaders as a means of dampening down the critical voices of the MPs.

In Finland the nuclear opposition inside Parliament has been paralysed along with the process that has obscured the relationship between the government and the opposition. Political opposition – in this case anti-nuclear opposition – cannot act as an efficient actor if it lacks the legal and political tools with which to hold the government accountable for its policies. The result has been that the members of the opposition camp – political parties, or individual politicians – have abandoned reform-oriented strategies, moderated their standpoints on the contested issue and tried to find channels and instruments with which to gain some policy benefits.

The role of the Finnish Greens in the nuclear policy processes of the early 2000s opens a perspective on the ideological moderation of the political parties in a state where power is centralized in the executive branch of the government. The Finnish Greens emerged in the late 1970s as a party that adopted a sharply negative position to nuclear energy. During the policy process of the early 2000s the Greens, however, participated in the government that gave a positive DiP to Parliament. The presence of the Greens in a government that had adopted a positive position in regard to the further building of nuclear energy was possible because the government's DiP was a majority decision. The Greens did not have to vote in favour of the pro-nuclear decision of the government. The Greens resigned from the government only after Parliament had approved the government's DiP in 2002.

The strategy of the Greens to stay in a government that – at least from the formal point of view – had made a pro-nuclear decision, attracted criticism from the general public. The Greens were accused of betraying their core principles. However, the Greens' strategy to stay in government also reveals the weakening power of Parliament as an institution through which the opposition could get its alternatives channelled to the government.

The strategies adopted by the politicians of the Centre Party also contribute to illustrating the tendency of political parties to moderate their political positions in an administrative state with a weak parliament. Traditionally, the Centre Party has tried to advance the use of domestic energy sources and to resist moves taken towards the construction of a centralized energy system. However, a large proportion of the representatives of the Centre Party have also become inclined to accept compromises between the nuclear energy line and the alternative line. More precisely, they have been willing to set as a prerequisite for their positive position on the further building of nuclear energy that the government makes a decision that shows its interest in the promotion of the use of domestic energy resources (Säynässalo 2005; see also the chapter of Lampinen in this book). The willingness of politicians to acquiesce to compromises is an aspect that largely explains why the majority of the members of the Finnish Parliament accepted the DiP of the government.

In my view the tendency of political parties to abdicate their responsibility to present and advance critical alternatives in the nuclear energy policy processes in France and Finland cannot be explained without attempting to explain the reasons why the political weakness of parties is considered a legitimate state of affairs, or at least natural or logical in these two countries. There is a need for an understanding of these two states as entities that provide a foundation for a particular kind of a state tradition. The centralized state structure with a centralized and a powerful administration is a feature of this tradition that describes especially the essence of the French state. Yet, as has been argued here, the French Constitution has also provided a foundation for the conception of a particular kind of democracy which can be used to legitimate a strong executive power. The question relates to the distribution of executive power to the parliament and to local governments in individual cases. This strategy, which has been conceptualized in this chapter as one of administrated democracy, was applied by the French Socialists in the early 1980s.

The concept of administrated democracy is of particular value when the logic of Finnish nuclear energy policy making is under consideration. The Nuclear Energy Act can be seen as a law that constructs a foundation for administrated democracy due to setting down a decision-making procedure that entrusts administrative authority to Parliament. This procedure has not provided a foundation for the emergence of disciplinary party/parties, such as the French Socialists in the early 1980s. The strength of the law derives rather from the fact that by setting down legal and administrative rules for decision making it has contributed to an erosion of the credibility of ideological politics in decision making concerning nuclear energy. This has been parallel to the growing willingness of politicians to moderate their anti-nuclear attitudes and to acquiesce to compromises.

The foundation for the legitimacy of the procedure set down by the Nuclear Energy Act derives from the same cultural tradition from which the French Fifth Republic draws its strengths. The broader issue is about a tradition in which more trust is placed in laws and legal procedures as a source of democratic governance than in party-based politics. It is not thereby an exaggeration to argue that the Nuclear Energy Act has constituted a foundation for the legal-administrative alliance between the state and the nuclear industry due to the advocation of the idea that nuclear companies have the legal right to apply for the DiP and the constitutional institutions have a legal responsibility to treat the application in accordance with the law, independently of the political balance of power.

Conclusions

The purpose of this chapter was to explain the particularities of the logic of nuclear policy making in Finland from a comparative perspective. The Finnish case can be regarded as something of a paradox because it has been possible for the Finnish government to advocate pro-nuclear politics, even though a particular law – the Nuclear Energy Act of 1987 – was specifically enacted to entrust Parliament with the authority to decide on the further construction of nuclear energy. However, the starting point of the chapter rested on the assumption that there are similarities between the logic of the nuclear policy making of Finland and that of France. I have also assumed that nuclear energy policy making in Sweden can be analysed as a policy process, the structural logic of which has been clearly different to that of the Finnish and French policy processes.

The choice to focus on these three states has been justified by way of viewing these states as representing the Continental European state tradition. This tradition has constituted a source of a particular kind of debate around the appropriate relationship between the state and democracy. I have argued that this tradition has provided a foundation for the emergence of two types of strong states in the postwar era: a strong party state and a strong administrative state.

Sweden has been analysed as an example of a strong party state. It has been characteristic for Sweden to allow the centralization of executive power in the reformist political party. This was in particular the case during the uninterrupted Social Democratic period of ascendancy from the early 1930s to the elections of 1976. The collapse of the Social Democratic government in 1976 reflected the vulnerability of party democracy in the face of new conflicts.

France has been analysed as a pure example of a strong administrative state. One characteristic of a strong administrative state has been to view the executive branch as a distinct public institution and thereby to detach it from parliament. The most well-known manifestation of a strong administrative state is the powerful bureaucracy of France that has been entrusted with strong and independent rule-making power. However, I have argued that a strong administrated state contributes to the provision of a foundation for the emergence of a phenomenon that has been conceptualized in this chapter as administrated democracy. The issue is about a form of democracy in which executive power is entrusted to parliament and possibly also to local institutions in individual cases. I have concluded that the purpose behind the application of administrated democracy is to legitimate the continuity of the chosen policy line rather than to provide influential channels for the social groups to gain influence in the state structures. It is argued that administrated democracy also offers a tool with which the politico-administrative elites are able to keep contested issues from strong ideological confrontations.

The brief empirical analysis focused on the nuclear energy policy processes of Sweden and France has contributed to a justification of the fertility of the distinction drawn between the two types of strong states. In Sweden the early development of nuclear energy became intertwined with the reformist politics of the governing Social Democratic Party. Consequently, the alliance between the state and the nuclear energy industry became openly political in character. This alliance was relatively easily challenged by the counter-reformist party, the Centre Party, during the mid-1970s. The electoral victory of the non-socialist opposition launched a process that led the Social Democrats to adopt an anti-nuclear position. In France, on the contrary, the powerful bureaucratic machinery played a crucial role in ensuring the continuity of the pro-nuclear policies. The strong role of the bureaucratic apparatus in ensuring the continuity of nuclear development provides a basis for arguing that the alliance between the state and the nuclear industry in France has been legal-administrative in character.

Finland cannot be regarded as a pure type of administrative state because the administration has not been equipped with the kind of authority as a leading force of economic and technological development as in France.

However, neither are the administrative-politico elites in Finland willing to leave economic and technological issues to be decided in the party-based parliamentary arena; not even when these issues become contested. This chapter has argued that Finnish constitutional practice (as well as that of France) provides tools for carrying out administrated democracy which contributes to preventing political parties from gaining a major role in decision making. It has been concluded further that the possibility provided by constitutional practice to entrust Parliament with an administrative authority in individual issues contributes to concentrating actual power instruments in the hands of administrative institutions. The decision making concerning the budget can be referred to as an example of administrated democracy. The budget procedure authorizes Parliament to make a final decision concerning the budget, but it is actually the Ministry of Finance that has the most important instruments of power at its disposal.

I have argued that the Nuclear Energy Act of 1987 has entrusted the Finnish Parliament with an administrative authority which means that the administrative task on individual issues is entrusted to Parliament. It has been concluded that this procedure actually entrusts a single ministry, the Ministry of Trade and Industry (since 1 January 2008 the Ministry of Employment and Economy) with powerful instruments to influence the course of the political decision making concerning nuclear energy.

I have also concluded that the Nuclear Energy Act has actually contributed to sustaining a legal-administrative alliance between the state and the nuclear industry in Finland. It has been the core of this alliance to assume that the nuclear companies have a legal right to apply for a DiP and the constitutional institutions (the government and parliament) have to treat the application in accordance with the formal procedure. Politics has therefore become subjugated to the legal-administrative procedure. This law has not contributed to providing a basis for the strict party discipline comparable to that carried out by the French Socialists in the early 1980s. Yet the formal procedure has contributed to a neutralizing of the political decision making from strong ideological confrontations. The nuclear energy debate in Finland has been intertwined with a complex interplay between party politics and constitutional politics. Consequently, politicians have become increasingly willing to moderate their anti-nuclear attitudes and to acquiesce into compromises that have made pro-nuclear politics possible.

Notes

1. In February 2009 the Swedish government decided to abandon its phase-out policy and to allow the construction of nuclear reactors in Sweden.
2. It has become a widely held view among social and political scientists that individual states and specific policy areas cannot easily or clearly be categorized (see e.g. Jasper 1990; Joppke 1992).

3. In the Anglo-Saxon tradition (to which, for example, the UK and the USA belong) there exists no corresponding awareness of the state as a political concept which builds into it a certain kind of collective identity, and as a legal institution with a responsibility for regulating matters of public concern (see Dyson 1980).
4. I am referring especially to the uninterrupted Social Democratic period of ascendancy from the early 1930s to the elections of 1976.
5. The concept of alliance in this connection is meant to refer to the close co-operation between the state and the nuclear industry in laying down the foundation for the growth of civilian nuclear technology (see, for example, Camilleri 1984).
6. The latter extremity comes from J.M. Jasper's book *Nuclear Politics: Energy and the State in the United States, Sweden, and France* (1990). This book is one of the most recent influential pieces of comparative research focused on nuclear energy policy. Although Jasper defines his book as one that does not aim to replace the structural perspective but aims rather to 'fill in' the structures, his book does not aim to develop institutional analysis, but instead largely brushes an institutional perspective aside.
7. The Constitution of the Fifth Republic granted the executive branch extensive powers independent of Parliament, as well as numerous opportunities to control parliamentary activity.
8. In Germany, the concept of the *Parteienstaat* is associated with the prominent role of political parties in the governance of the state, behind which has been a conscious effort to make the state more penetrable for civil society interests. This effort has been associated with giving the parties a constitutional mandate to participate in the formation of the people's political will with the purpose of supporting the presence of partisans in the key positions of the state apparatus (Dyson 1980; Joppke 1992).
9. The most visible one was the decision of the government in 1948 to establish a special organization – AB Atomenergi – to coordinate all research activities and safeguard state control over nuclear development. The government also appointed a new Atomic Commission in 1955 to draw up plans for the future development of nuclear energy. The plans of the established commission gave expression to the strong commitment of the government to the development of nuclear power as a substitute for imported oil and as a means of gaining self-sufficiency in electricity production (for more detailed information, see Kaijser 1992, 443–6).
10. The collapse of the 'Swedish Line' was caused by many different factors. One of the main ones was the new knowledge available at the end of the 1950s which indicated that the costs of new reactors were higher than the Commission had estimated. Secondly, there was a hidden military aspect inherent in the ambitious programme, which was an aspect that divided the Social Democrats in the early 1960s. The decision made by Parliament in 1968, according to which Sweden would not develop nuclear weapons, reduced the feasibility and legitimacy of the 'Swedish Line'. Thirdly, the plans of AB to build a reactor in Marviken met with complications in 1957 (see Kaijser 1992, 446).
11. In the area of the nuclear industry the interventionist line of the Social Democrats led to the emergence of a new company called AB ASEA-Atom (a combination of AB and ASEA) with ASEA and the Swedish state as 50/50 owners (see Kaijser 1992).
12. By the end of 1971 four of Sweden's nuclear sites had received construction permits for at least one reactor and eight reactors in total had been licensed for construction.
13. The constitutional reform replaced a bicameral parliament with a unicameral one. The result was that the removal of the second chamber was that the Social

Democrats found it increasingly difficult to sustain a majority in Parliament (see Pontusson 1992; Arter 1999).

14. These reforms were justified by the view according to which parliamentary majorities should not be hindered from conducting substantial policies of reform which involved government intervention in a variety of social fields (see Algotsson 2001).
15. The cornerstone of the anti-nuclear turn was the result of the 1980 referendum.
16. At the end of the 1960s EDF (Électricité de France) managed to win the battle over reactor technologies. EDF advocated the so-called light-water reactor-type (LWR) while CEA (Commissariat à l'Énergie Atomique), which had influenced the early stage of the French nuclear sector, was committed to the gas-graphite reactor-type. EDF won the battle as the result of growing political support for the LWR from President George Pompidou who supported the argument promoted by EDF according to which the LWR strengthened the economic competitiveness of the nuclear industry (see Jasper 1990).
17. This Commission was established to advise CEA in its effort to develop nuclear reactors capable of producing electricity. The members of this Commission consisted of representatives from CEA, EDF, the state and industry. It occupied an important role in giving formal recommendations to the adoption of the line of nuclear policy favoured by CEA or EDF (Jasper 1990).
18. The shift that was taken in energy policy represented almost 13,000 MWe. By the year 2000, nuclear energy was intended to account for 50 per cent of France's total energy needs (see Jasper 1990, 156).
19. The Finance Ministry was perhaps the most powerful actor among all those who criticized the Messner Plan. From the Finance Ministry's point of view, the plan was too costly for the state (see Jasper 1990).
20. The French state had focused repressive strategies against the participants of a demonstration that was organized in Creys-Melville in 1976. The demonstration was targeted against the Super Phoenix breeder reactor that was proposed for this site (for more detailed information, see Jasper 1990).
21. The question was about annulling a report critical of the large nuclear programme. This report was composed by the group of the Assembly headed by the Socialist deputy Paul Quilès, who had been allowed to publish a report critical of the nuclear programme just before the 1981 elections. After the elections the Socialist party leaders were more willing to support the so-called Hugon Report which favoured the expansion of nuclear energy. The Prime Minister threatened to resign if Quilès opposed the Hugon Report in the voting in the Assembly. Even though there had been strong anti-nuclear attitudes among the Socialist deputies before the voting on the Hugon Report, it passed easily after two days of debate (see Jasper 1990, 247–8).
22. This Constitution replaced the first Constitution of independent Finland (1919).
23. A postponement law could come into reconsideration in the first session of Parliament after the next general elections and it would come into force if it was supported by the majority of MPs.
24. The position of political parties, especially left-wing parties, was problematic after the Civil War of 1918 – a war that divided Finnish society into two camps. The strong ideological tensions between the left- and right-wing forces were a visible element of political life from 1918 until the late 1970s.
25. These two governments led by Lipponen of the Social Democratic Party (the first was founded in 1995 and the second in 1999) have been labelled 'rainbow

coalitions'. The Social Democrats, together with the National Coalition Party, constituted the main government parties in these regimes. The other government parties were the Left Socialists (Left Alliance), the Greens and the Swedish People's Party.

26. IVO was forced to order the first set of nuclear reactors from the Soviet Union.
27. The president's support was important for TVO because the Social Democrats who belonged to the socialist–agrarian majority government (appointed in 1972) were opposed to the plan. The Social Democrats regarded the plan as one that contributed to giving birth to the private nuclear plants which were against the official line of the party. Yet TVO's plan turned out to be a complex one for the Social Democrats with the result that the socialist–agrarian government was forced to accept TVO's plan. TVO ordered the second set of NPPs from the Swedish Asea-Atom (see Sunell 2001).
28. For more about the attempts of the nuclear industry to advance a contract for building the fifth nuclear reactor, see Michelsen and Särkikoski (2005).
29. In 1976 the government appointed a committee to compile a proposal for reforming nuclear legislation.
30. For example, the nuclear legislations of USA, France and Sweden have included stipulations concerning public hearings and/or veto power granted to municipalities concerning nuclear sites (see Jasper 1990).
31. As is the case when the budget power of Parliament is under evaluation.
32. The decision to vest Parliament with the administrative authority by the Nuclear Energy Act was considered to be quite exceptional in the Finnish political and constitutional culture. Yet the Nuclear Energy Act was enacted as an Exceptive Act. The enactment of Exceptive Acts has been quite commonplace in the Finnish constitutional culture (see Kasurinen 2000; Jyränki 2003).
33. The paradox is that the Finnish Parliament itself constitutes a central arena for and a focal point of constitutional politics. Parliament is responsible for the reviewing of the constitutionality of Parliament Acts. This system of preventive control has been the task of the Constitutional Committee of Parliament which is consisted of ordinary MPs. It has been one of the main tasks of this committee to give recommendations on the proper interpretation of the Constitution in particular cases (see e.g Jyränki 1989; Länsineva 1991).
34. When the government bill was under discussion in Parliament, the MPs suggested different possibilities for deciding on the use of nuclear energy. It was suggested, for example, that the decision made by Parliament on the construction of the new nuclear reactors could take the form of law which would provide Parliament with a strong power instrument in nuclear energy policy. The suggestion to submit the decision concerning the construction of a new reactor to be solved in the referendum was also raised up (see Ruostetsaari 1986, 171).
35. The company was established as a cooperation between TVO and IVO to advance the plan for the fifth nuclear reactor.
36. The question was about taking a stance on the report of the Economy Committee (26/1992). The report touched on the Account of the Finnish energy strategy drawn up by the government. The report did not take a position on the further building of nuclear power plants; instead, it raised different alternatives for Finland's energy strategy. However, the resolution that became connected to it included a negative stance to the further building of nuclear power. This resolution that was initiated by the current Prime Minister Matti Vanhanen (who was an MP at this time) was voted on and it became accepted by a clear vote – 96–78.

37. In this light labelling the nuclear energy issue as one of conscience can be readily seen as an act that supported the goals of the anti-nuclear camp because of justifying the moral arguments and free discretion of politicians in the deliberation of nuclear energy.

References

Algotsson, K.-G. (2001) From Majoritarian Democracy to Vertical Separation of Powers: Sweden and the European Union. *Scandinavian Political Studies* 1, 51–65.

Almond, G. (1988) The Return of the State. *American Political Science Review* 82, 853–74.

Arter, D. (1999) *Scandinavian Politics Today*. Manchester: Manchester University Press.

Atomienergialaki 356/1955. (Atomic Energy Act of 1955) Suomen säädöskokoelma.

Camilleri, J.A. (1984) *The State and Nuclear Power: Conflict and Control in the Western World*. Brighton: The Harvester Press Ltd.

Campbell, J.L. (1988) *Collapse of Industry: Nuclear Power and the Contradictions of US Policy*. Ithaca, NY: Cornell University Press.

Cerny, P.G. (1990) *The Changing Architecture of Politics: Structure, Agency, and the Future of the State*. London: Sage.

Dyson, K. (1980) *The State Tradition in Western Europe*. New York: Oxford University Press.

Evans, P., D. Rueschemeyer and T. Skocpol (eds) (1985) *Bringing the State Back In*. New York: Cambridge University Press.

Hallituksen esitys ydinenergialaiksi 16/1985 (The Government Bill for the Nuclear Energy Act of 1985). Asiakirjat A1. Hallituksen esitykset.

Harrinvirta, M. (1997) Miksi ja miten budjettia perustellaan? Perustelukäytäntöjen merkitys, muutos ja ongelmat (Why and How is the Budget Being Justified? The Meaning, Change and Problems of the Practices of Justification). *Hallinnon tutkimus* 2, 153–60.

Hecht, G. (1998) *The Radiance of France: Nuclear Power and National Identity after World War II*. London: The MIT Press.

Jasper, J.M. (1990) *Nuclear Politics: Energy and the State in the United States, Sweden and France*. Princeton, NJ: Princeton University Press.

Joppke, C. (1992) Models of Statehood in the German Nuclear Energy Debate. *Comparative Political Studies* 2, 251–80.

Jyränki, A. (1989) *Lakien laki. Perustuslaki ja sen sitovuus eurooppalaisessa ja pohjois-amerikkalaisessa oikeusajattelussa suurten vallankumousten kaudelta toiseen maailmansotaan* (The law above all laws. Constitutional law and its validity in the European and North-American legal thinking from the era of the grand revolutions to the Second World War). Helsinki: Lakimiesliiton Kustannus.

Jyränki, A. (2003) *Valta ja vapaus. Kaksikymmentäkolme luentoa valtiosääntöoikeuden yleisistä kysymyksistä* (Power and Freedom. Twenty-three Lectures on the General Issues of Constitutional Law). Helsinki: Kauppakaari Oy.

Kaijser, A. (1992) Redirecting Power: Swedish Nuclear Power Policies in Historical Perspective. *Annual Review of Energy Environment* 17, 437–62.

Karjalainen, K. (1989) *Politiikka, talous ja energiatalouden poliittinen ohjaus Suomessa* (Politics, Economy and the Political Steering of the Energy Economics in Finland). Imatran Voima Oy Tutkimusraportteja 13.

Karjalainen, K. (1996) Suomen energiapolitiikasta (On the Finnish Energy Policy). Imatran Voima Oy. Tutkimus- ja kehitysyksikkö. Raportti, Sira-A9-6.

Kasurinen, N. (2000) *Ennen 1.3.2000 säädettyjen poikkeuslakien suhde uuteen perustuslakiin* (The Relationship of the Exceptive Laws Enacted before 1st March 2000 to the new Constitution Act). Oikeusministeriön selvityksiä (http://www.om.fi/uploads/hyfju3doe9.pdf).

Katzenstein, D. (ed.) (1978) *Between Power and Plenty*. Madison, WI: University of Wisconsin Press.

Kesselman, M. (1992) France. In K. Kesselman et al. (eds), *European Politics in Transition*. Lexington, MA: D.C. Heath and Company.

Krasner, S. (1984) Approaches to the State: Alternatives, Conceptions and Historical Dynamics. *Comparative Politics* 16, 223–46.

Kyllönen, S. (2004) Ydinvoiman ilmastonmuutos (The Climate Change of Nuclear Power). In M. Kojo (ed.), *Ydinvoima, valta ja vastarinta* (Nuclear Power, Power and Resistance). Helsinki: Like, pp. 51–85.

Länsineva, P. (1991) The Constitutional Committee of Parliament: The Finnish Model of Norm Control. In M. Sakslin (ed.), *The Finnish Constitution in Transition*. Helsinki: The Finnish Society of Constitutional Law.

Lindvall. J. and B. Rothstein (2006) The Fall of the Strong State. *Scandinavian Political Studies* 1, 47–63.

Michelsen, K-E and T. Särkikoski (2005) *Suomalainen ydinvoimalaitos* (The Finnish nuclear power plant). Helsinki: Edita.

Nelkin, D. and M. Pollak, (1980) Political Parties and the Nuclear Energy Debate in France and Germany. *Comparative Politics* 2, 127–41.

Nousiainen, J. (2001) From Semi-presidentialism to Parliamentary Government. Political and Constitutional Developments in Finland. *Scandinavian Political Studies* 2, 95–109.

Paloheimo, H. (2003) The Rising Power of the Prime Minister in Finland. *Scandinavian Political Studies* 3, 219–43.

Pontusson, J. (1992) Sweden. In K. Kesselman et al. (eds), *European Politics in Transition*. Lexington, MA: D.C. Heath and Company.

Richardson, Jeremy (ed.) (1982) *Policy Styles in Western Europe*. London: George Allen & Unwin.

Rockman, Bert A. (1990) Minding a State – Or a State of Mind? Issues in the Comparative Conceptualization of the State. *Comparative Political Studies* 23, 25–55.

Ruin, O. (1982) Sweden in the 1970s: Policy-making Becomes More Difficult. In J. Richardson (ed.), *Policy Styles in Western Europe*. London: George Allen & Unwin.

Ruostetsaari, I. (1986) *Energiapolitiikan päätöksenteon ja hallinnon kehityksestä ja nykytilasta Suomessa* (On the Development and the Present State of the Decision-making and Administration of Energy Policy in Finland). University of Tampere, Research Reports 85.

Ruostetsaari, I. (1989) *Energiapolitiikan määräytyminen* (The Determination of Energy Policy). Acta Universitatis Tamperensis, ser A, vol 278. University of Tampere, Tampere.

Skocpol, T. (1979) *States and Social Revolutions: A Comparative Analysis of France, Russia and China*. Cambridge: Cambridge University Press.

Sunell, M. (2001) *Miten Suomen yksityinen metsäteollisuus hankki länsimaisen ydinvoimalan. Tutkimus taloudellisesta ja poliittisesta vallankäytöstä 1970-luvulla* (How the Finnish Private Forest Industry Acquired a Western Nuclear Power Plant. A Study on Use of Economic and Political Power in the 1970s). University of Helsinki. Department of Social Science History. Unpublished Masters thesis study. http://ethesis.helsinki.fi/julkaisut/val/yhtei/pg/sunell/mitensuo.pdf.

Sunell, M. (2004) Suomalainen ydinvoimapoikkeus (The Finnish Nuclear Energy Exception). In M. Kojo (ed.), *Ydinvoima, valta ja vastarinta* (Nuclear Power, Power and Resistance). Helsinki: Like.

Suomen perustuslaki (2000) (The Constitution Act of Finland of 2000). Helsinki: Eduskunta.

Säynässalo, E. (2005) Between Politics and Law: Explaining the Failure of the Greens to Prevent the Further Building of Nuclear Energy in Finland. Paper presented at XXXVIIIth Annual Conference of Finnish Political Scientists, University of Jyväskylä, 12–13 January 2006.

Talousvaliokunnan mietintö 26/1992. (Report of the Economic Committee of 1992) Asiakirjat: Valiokuntien mietinnöt ja lausunnot C2.

Tiihonen, P. (1990) Legal, Political and Practical Aspects of Budgetary Decision-making in Finland. *Scandinavian Political Studies* 44, 327–46.

Tiihonen, P. and S. Tiihonen (1990) *Keskitetyn suunnittelun nousu ja tuho. Jäähyväiset alistavalle pysähtyneisyyden ajan suunnittelulle* (The Rise and Collapse of Centralized Planning. A Farewell to the Repressive Planning of the Era of Stagnation). Helsinki: Valtion Painatuskeskus.

Tuori, K. (2005) Vallanjako – vaiettu oppi (Separation of Powers – the Silent Doctrine). *Lakimies* 7–8, 1021–49.

Ydinenergialaki 990/1987 (Nuclear Energy Act of 1987). Helsinki: Suomen Säädöskokoelma.

Part IV
Nuclear Waste and Societal Risks

6

The Strategy of Site Selection for the Spent Nuclear Fuel Repository in Finland

Matti Kojo

Introduction

Posiva Ltd (Posiva), the company responsible for practical preparations and implementation for the final disposal of spent nuclear fuel (SNF) in Finland, suggested that the Government of Finland should consider only Olkiluoto in Eurajoki in its application of a Decision-in-Principle (DiP) for the final disposal site in May 1999 (Posiva 1999a, 5–6). The municipal council of Eurajoki approved a favourable statement on the Decision-in-Principle on 24 January 2000. The government made the DiP on 21 December 2000, and Parliament ratified the decision on 18 May 2001.

The aim of this chapter is to analyse the site selection process for an SNF repository in Finland. The main question under consideration will be why the Olkiluoto area in the municipality of Eurajoki was chosen. The structure of the chapter considers three subtasks that have been derived from the concept of the siting strategy. The first subtask is to offer a brief clarification of the development of nuclear waste policy in Finland. The chapter describes how final disposal was chosen to be the main alternative in Finland, despite the fact that the transportation and reprocessing of SNF abroad had been the starting point of Finnish policy until the early 1990s. Thus this part of the chapter examines how the siting process was framed by the nuclear waste policy. The second subtask is to give an overview of the changes in the site selection process since the early 1980s. The focus is in particular on geological and non-geological siting criteria. The third subtask is to study local decision making in Eurajoki, and in particular the discussions between the municipality of Eurajoki and the nuclear power company Teollisuuden Voima Ltd (TVO) and later Posiva. This part focuses on the development of a partnership between the main local players.

These three subtasks form a picture in which nuclear waste policy frames the formal site selection process, whereas the analysis of local decision making reveals more about the informal side of site selection. The chapter concludes that the long-term interaction between the representatives of the candidate municipality and the nuclear power company TVO was the crucial factor in the decision-making process. Decision making in Eurajoki is analysed as a case study, in which the parties reached an agreement on the compensation for the final disposal repository. The negotiations between the municipality of Eurajoki and the nuclear energy industry involved in reaching a positive decision are analysed from the beginning of the 1980s. The main emphasis, however, is on the years 1996–99, when the nuclear energy industry negotiated with the municipality on the compensation for the final disposal repository.

According to Sundqvist (2002, 110),

> a site selection strategy is the base from which the surrounding world is interpreted, and also identifies the tasks that have to be carried out. The strategy is used as a tool for understanding, interpreting and manipulating reality, and will therefore shape the identity of the organisation as well as its view of the external world.

Sundqvist (2002, 110) states that the formation of a siting strategy means developing knowledge and, simultaneously, developing relationships with the actors that one has to work with, not least the citizens and politicians in the municipalities involved in the siting process. The concept of a site selection strategy is applied in the chapter as an analytical tool.

Sundqvist's concept is based on an analysis of Swedish nuclear waste policy from which the technological core, the KBS 3 geological disposal method, is applied to the case of Finland. Sundqvist describes two different strategies. In a systematic strategy geological siting criteria are emphasized, whereas in a flexible strategy the emphasis is on voluntariness and local acceptance (Sundqvist 2002, 114–16). According to Anttila (1995, 7), the elimination of potential sites was based on purely geological criteria in Finland and, thus, siting followed a classical elimination process (Richardson 1998, 10). By the mid-1990s both Litmanen (1994, 23, 139–41) and Anttila (1995) had already concluded that the purely geological elimination process was in change in Finland and that environmental and social criteria were being emphasized rather than purely geological criteria.

Deviation from the systematic siting strategy happened in the early stages of the research. Litmanen (1994, 23) comments that geological criteria were applied to the selection of the areas, but that the investigation sites were not chosen on strictly geological principles. Anttila (1995) repeats this argument, stating that in the selection of the final disposal site the importance of environmental and social factors has clearly overtaken geological criteria in recent years. As regards different approaches to site selection, Anttila distinguishes between a direct geological selection and an approach based on the voluntariness of the municipality.

The structure of this chapter is as follows: The second section highlights the main milestones of nuclear waste policy development in Finland. The third section is focused on changes in the site selection process. The fourth section provides background on how nuclear waste became a political issue in Eurajoki in the 1970s and 1980s. The fifth section describes how co-operation between the main players became closer in the 1990s. The sixth section focuses on the debate on the compensation agreement negotiated between the municipality, TVO and Posiva. The final section provides some overall conclusions.

The nuclear waste policy in Finland

The current Finnish nuclear power programme includes four nuclear power plant (NPP) units that were built in the 1970s and early 1980s. Two of these are situated in Loviisa one hundred kilometres east of the capital Helsinki and two in Eurajoki some 220 km north-west of Helsinki. The NPP in Loviisa is operated by Fortum Power and Heat Ltd (FPH), which is a part of the Fortum Consortium. FPH is the former Imatran Voima Ltd (IVO). The company was renamed in the late 1990s. The main owner of Fortum is the Finnish state. The other nuclear power company in Finland is Teollisuuden Voima Ltd, which operates the NPP in Eurajoki. TVO is owned by Pohjolan Voima (PVO), which in turn is owned by heavy industry interests such as forest companies. Fortum also owns one-quarter of TVO. In November 2000 TVO applied for a DiP for a new NPP (see Appendix 1 at the end of chapter 8). After a period of heated public debate Parliament ratified the DiP in May 2002 (see Chapter 2 by Lampinen in this book; see also Lammi 2004). The new 1,600 MW NPP unit is under construction in Eurajoki. According to the latest timetable, the unit should be operational in 2012 – after a delay of three years.

The starting point for the repository siting process was formulated along with the overall nuclear waste policy at the end of the 1970s and the beginning of the 1980s. Finnish nuclear waste policy was originally based on reprocessing, because in the 1960s the level of known uranium sources was still low. Earlier, in the 1960–70s, when the NPPs were ordered, the question of which countries it was suitable for Finland to import SNF from was politically more important than SNF management (Suominen 1999, 25; Sunell 2004, 186). There was a fairly optimistic attitude to SNF, as it was thought that there was a use and a market for it.

In the beginning, TVO alone had to decide how to manage the SNF produced by the NPP it owned. IVO returned the spent nuclear fuel originating from Loviisa NPP to the Soviet Union, and later to Russia, during the years 1981–96. IVO's SNF management was based on the fact that the high-level waste created from reprocessing should stay in the Soviet Union. The return was based on the agreement signed between the governments of Finland and the Soviet Union in 1969 regarding the use of nuclear energy in peacetime. This arrangement corresponded to the principle that the Soviet

Union undertook to deliver the fuel and receive the resulting SNF from those reactors which had been delivered to other countries (Sandberg 1999, 45–6).

In Autumn 1976 the MTI established a working party on active waste and fuel (APO) to consider questions of responsibility, and an organization to deal with nuclear waste management and questions dealing with the development of nuclear energy legislation. In March 1978 the statement of the working party to the MTI laid out the general outlines for alternative nuclear waste management. In the statement, reprocessing in Finland was not considered a reasonable alternative, and therefore SNF should be exported. The statement also stressed the long-term storage of SNF, and its possible final disposal in Finland (Suominen 1999, 25). The main alternatives in nuclear waste management were therefore as follows: a long-term storage or direct final disposal facility in Finland, and the reprocessing and final disposal abroad.

The government made a DiP regarding the arrangements for nuclear waste management in April 1978 based on the statement by the APO working party. In accordance with the DiP, the execution of the preparations for nuclear waste management is the duty of the producers of the waste, that is, the nuclear power companies, who are also responsible for the costs.

The government's DiP was applied to the licences granted to TVO and IVO. The TVO I licence was issued on 6 July 1978. The government's prerequisite was that the applicants for the licence must attempt to export the SNF from Finland. According to Björklund, Westerholm and von Bonsdorff (1994, 145), the Finnish licence authorities stressed the importance of reprocessing, as similar conditions were set for the licences of NPPs in other countries. However, the licence issued to TVO had a provision requiring TVO to present a plan for the final disposal of SNF in Finland to the MTI. Consequently, the licence granted by the authorities was not as restrictive regarding the establishment of the reprocessing agreement as the corresponding licence in Sweden.[1] The management of TVO, however, had different opinions about how important the authorities regarded the reprocessing agreement included in the licence (Björklund et al. 1994, 145, 147–9). The licences also contained an obligation to carry out research and planning. Thus by the end of the 1970s there were already some plans for possible final disposal in Finland (Suominen 1999, 26), but at this time final disposal was not the primary alternative.

In 1973 TVO conducted the first negotiations regarding reprocessing. At that time there was still some excess capacity in global reprocessing, but this situation changed towards the end of the 1970s. In 1975–76 the TVO waste team made a presentation of various alternatives, one of which was long-term temporary storage, and later direct final disposal. This alternative then became practicable as the availability of uranium improved and the price dropped. At the same time the cost of reprocessing rose, and the conditions tightened. For example, under the new conditions high-level waste left over from reprocessing should have been returned to the country of origin, and the customers should have been responsible for the financing of the entire construction of the repository through advance payments by means of an interest-free loan (Björklund et al. 1994, 145–6). The negotiations regarding the reprocessing agreement with the French company COGEMA and the British company British Nuclear Fuels Ltd (BNFL) did not result in the signing of an agreement, because the vice-chairman of the TVO board, Wolter Westerholm, specifically objected because of the high costs involved. The latter reprocessing tender negotiated with BNFL was rejected by the board of TVO in December 1979. TVO felt confident in rejecting the BNFL tender having learned of MTI's attitude to the matter. The MTI did not consider that the rejection of the tender infringed the conditions of the TVO licence (Björklund et al. 1994, 147–9).

Although reprocessing was the primary alternative to TVO's nuclear waste management plans in Finland's nuclear waste policy, TVO had the courage to dismiss reprocessing tenders. The reasons were financial, from the point of view of both the company and the national economy (Björklund et al. 1994, 149). On the other hand reprocessing was outlined as an alternative in the TVO research reports. The obvious attempt here was to present TVO to the authorities as an active and voluntary observer of the licensing conditions (for example, Raumolin 1982, 5). The final disposal was to be carried out in a concentrated fashion abroad, since 'both from the point of view of safety and economics it would obviously be more profitable, if the temporary storage and final disposal of the waste was carried out in a concentrated way, e.g. in the countries being major producers of nuclear energy' (Ryhänen 1979, 11, 12). However, the evaluation of the nuclear waste management situation changed rapidly; in September 1981 TVO's most important objective was to acquire more storage capacity, allowing more time to follow the international development and evaluation of the alternatives in nuclear waste management (Ryhänen et al. 1981, 42–3). The company arrived at this conclusion because SNF management was an open question globally. The company maintained it would be sensible for Finland to wait until matters had become clearer elsewhere. The company was also prepared to store spent fuel in the power station's pools until the end of the 1980s (Björklund et al. 1994, 146). According to Suominen (1999, 29), the rationale behind this was the change in the international political situation. The reasons given by Björklund, Westerholm and von Bonsdorff (1994) regarding the company's finances, however, seemed to be the primary reason for dropping the reprocessing alternative. According to TVO's deputy managing director, Haapala (1988, 10), the choice of the reprocessing alternative was inhibited by the costs, the shortage of reprocessing capacity and the fact that the waste management was left to the waste producers. In the early 1980s TVO was also offered the possibility of signing an agreement on the interim storage of spent fuel in Sweden. According to Haapala (1988, 10), the building of their own interim storage was preferred because of the timetable and the costs involved.

Research on the safety of SNF management and the final disposal began in 1977, when the International Atomic Energy Agency (IAEA) instigated an international nuclear fuel reprocessing evaluation (the INFCE Project, International Nuclear Fuel Cycle Evaluation). At the beginning the Ministry of Trade and Industry in Finland financed the research, which was based as a scientific institution. In accordance with operating licences, the nuclear energy companies Teollisuuden Voima and Imatran Voima established the Power Companies' Nuclear Waste Committee (Voimayhtiöiden ydinjätetoimikunta YJT) in 1978. The task of the Nuclear Waste Committee was to plan and lead research work in the field of nuclear waste management. It set the research programmes for the years 1978–88 and was responsible for their publication. During the period 1989–2001 three research programmes – called the JYT research programmes – were carried out, funded and coordinated by the public authorities (Rasilainen and Vuori 1999, 42–3; Suominen 1999, 25). Since 2002 research has been carried out under the Finnish Research Programme on Nuclear Waste Management (Kansallinen ydinjätetutkimusohjelma KYT). The Nuclear Energy Act was amended in late 2003 to ensure funding for research into long-term nuclear safety and nuclear waste management. The finance is collected annually from the licence holders – nuclear power companies – in two special funds (MTI 2005, 19–20).

In order to promote the research and planning work in 1978 the MTI founded a nuclear waste working party, whose task was to draft a proposition for the guidelines and schedule of research into nuclear waste. The statement produced by the working party in February 1980 consisted chiefly of a timetable for TVO regarding the implementation of the schedule of nuclear waste management. The statement included two alternatives: SNF management should take place either abroad or in Finland. Transportation abroad irrevocably for reprocessing or disposal was the primary alternative, in respect of which the working party proposed that TVO should reach an agreement on final disposal taking place abroad by 1983. Should this main alternative not materialize, Finland would then start preliminary studies on final disposal on its own territory (Suominen 1999, 26–7). As stated before, in 1979 the board of TVO – at the request of Westerholm – rejected the negotiation of reprocessing tenders.

In 1983 the Government of Finland made a decision according to which the primary alternative was now the storage of reprocessed high-level nuclear waste or SNF abroad. The reason given was the small amount of SNF produced in Finland. IVO's SNF management was already in line with the policy of the government since in 1981 the company had transported SNF to the Soviet Union for the first time. The DiP did, however, demand that the licence holders TVO and IVO should be prepared for final disposal in Finland if necessary. The demand concerned the disposal of such spent fuel as was not transported for permanent storage abroad. Thus, the governmental decision left the door open for the disposal alternative in Finland. This was especially important for

Table 6.1 Timetable for spent fuel final disposal

1980–1982	Suitability study with safety analyses
1983–1985	Preparation for the preliminary site characterisation
1986–1992	Preliminary site characterisation in chosen areas (5–10 sites)
1993–2000	Additional siting studies (2–3 sites)
2001–2010	Detailed studies of chosen disposal site and preplanning of the siting and the encapsulation plant
2011–2020	Planning and construction of the disposal site and the encapsulation plant
2021–2050	Final disposal facility is operational
2050–2060	Closing of disposal site

Source: Raumolin (1982, 7).

the TVO board, which had rejected negotiations on reprocessing tenders as being too expensive. As a consequence, international discussions on reprocessing, and the rise in costs, had a marked effect on Finnish nuclear policy.

The government's 1983 decision also included the schedule for nuclear waste management. The research programme consisted of three sections: (1) before the end of 1985 site identification surveys for the choice of investigation areas; (2) before the end of 1992 preliminary site characterization; and (3) before the end of 2000 detailed site characterization in those areas which had been shown to be most suitable at the previous stage (Council of State 1983). The government's timetable was based on the schedule presented in the TVO programme (Raumolin 1982, 5, 7) for the final disposal of spent fuel (Table 6.1). As a basis for the schedule it was suggested that the fuel should be allowed to cool for 40 years prior to its final disposal. During that period the activity and heat production of the SNF would drop to just 10 per cent of what they were only a year after being taken from the reactor.

The schedule, which directly concerns only those responsible for nuclear waste management, has been changed on only one occasion. In October 2003 the MTI decided that the nuclear power companies were now obliged to present their final plans for the application of the construction plan of the repository no later than 2012. Originally, it was intended that this should happen by 2010. The preliminary plans must be submitted in 2009. The reason for the change was to safeguard the safety of final disposal (Kojo 2004b, 232).

IVO's spent fuel transports to Russia ended in 1996, when the export ban on nuclear waste included in the Nuclear Energy Act came into force at the same time as the expiry of the return agreement (Sandberg 1999, 50–1, 58–9). The export ban was opposed by both IVO and the Radiation and Nuclear Safety Authority. The working party evaluating Finland's nuclear waste programme concluded in its 1988 report that because of changed market forces and the contractual obligations for the return of waste from reprocessing, such treatment did not seem 'at least on a short-term basis to be profitable from a technical-economic point of view' (MTI 1988, 16–17). The Nuclear Energy Act did not regard waste transport abroad as preferable to its final disposal in Finland. The working party considered it to be sensible, even if the export objective was 'understandable, it cannot always be explained from any other standpoint than owing to selfish benefits to the waste producing country' (MTI 1988, 28). The working party did not, however, regard direct final disposal as the only future alternative, and recommended a readiness to face a situation when the use of foreign services would be more acceptable than direct final disposal at home. This was in spite of the fact that the former would be a great deal more expensive than the latter (MTI 1988, 35). TVO did not abandon the reprocessing alternative based on the decision by the MTI regarding the objectives of nuclear waste management until 1991. For IVO, the same principles of nuclear waste management came into force following MTI's decisions in 1995.[2] The nuclear power companies agreed on the establishment of a joint nuclear waste company in 1995, which crystallized the concentration of a national nuclear waste management strategy as included in the altered Nuclear Energy Act. Before that, IVO had contemplated establishing its own final disposal repository (Anttila 1995), clearly a tactical move concerning inter-company negotiations.

Pragmatic siting strategy

Bedrock studies on the final disposal of SNF have been carried out in Finland since the latter part of the 1970s. During the period 1980–82 studies were carried out on the general suitability of Finland's bedrock for final disposal and of macro structures. In November 1982 TVO presented a conclusion maintaining: 'that surveys conducted in Finland and abroad in areas with similar bedrock show that suitable geological formations for the final disposal of spent nuclear fuel can be found in many areas in Finland' (Final disposal... 1982, 121). The summary of the report stated that surveys indicate that spent nuclear fuel can be disposed of safely in the Finnish bedrock (Final disposal... 1982, Summary). Geology was presented as the main criterion for site selection, but no detailed criteria were given. Regarding site selection, it was concluded that the decision would be based on the information obtained from the site selection studies, and that the final disposal site of SNF would be chosen on the basis of detailed site characterization (Final disposal... 1982, 31, 36).

The starting points of TVO's nuclear waste management operation were as stated in the programme published in December 1982: universality, known technology, safety, right timing and flexibility (Raumolin 1982, 1–2). Flexibility entailed being prepared for changes in the operation times of the NPP units and the amounts of waste arising as well as changes in the management methods. TVO's selection principle for investigation areas was that 'first geological studies are carried out and only then will those sites which are geologically suitable be chosen from the areas which are also suitable as research areas because of other aspects than from a geological angle' (Ryhänen 1985, 32). In this manner TVO applied a siting strategy which combined different selection principles. The strategy was therefore not based solely on geological factors.

In 1983 TVO activated its research and development programme, which also included measures for site selection. From a geological perspective the studies were initially systematic, when the whole of Finland was being researched in area selection studies between 1983 and 1985. This was an ideal model, where the selection of investigation areas was to be based on step-by-step phased studies (Litmanen 1994, 23). According to McEwen and Äikäs (2000, 33), the approach taken by TVO to site selection was, initially, to survey the whole of Finland, and then, from the data collected, to try to exclude areas as being unsuitable for repository development based on simple geological reasoning. TVO's philosophy was to focus quickly on regions that looked promising and not to waste time investigating large parts of the country. TVO commissioned studies on siting from the Geological Research Centre and also from the consultancy company Saanio & Laine. In the period from 1983 to 1985 target areas of the bedrock were given a preliminary examination to establish the general site characteristics. The target areas were vast, 100–200 km^2, taking into account the fact that the area needed for the final disposal was estimated as 40 hectares, i.e. 0.4 km^2. In total some 327 such target areas were mapped by the Geological Research Centre, based on geological criteria.

On the basis of environmental factors the target areas were narrowed down from 327 to 162, and were then narrowed down further, on geological principles, to 61 target areas. The environmental factors taken into account were population density, transport-related factors, land-use planning restrictions and protected areas and important groundwater resources. The principal criteria for further reduction were the size of the target area, the probability of locating smaller bedrock expanses within the target area, the fracture density, topography and the degree of geological exposure (McEwen and Äikäs 2000, 35–43).

In these 61 target areas, 134 investigation areas were identified. When choosing suitable bedrock, the general principle was to choose the most solid rock blocks based on the information obtained from the bedrock research, the so-called 'block-within-a-block' principle. This principle was based on the assumption that the blocks within the investigation areas would remain intact during any future deformation of the rock mass (McEwen and Äikäs 2000, 41). The selected investigation areas were already considerably smaller in size than the target areas, which covered hundreds of square kilometres, but the investigation areas still covered several square kilometres (5–10 km^2).

These 134 possible investigation areas were selected for further analysis. Based on the overall geological evaluation carried out by the Geological Research Centre, the investigation areas were classified into four geological suitability classes. In parallel with the geological classification, TVO divided the areas into three classes according to non-geological factors (population density, transportation conditions and land ownership). Thus the investigation areas were ranked according to preference (McEwen and Äikäs 2000, 43–6). According to McEwen and Äikäs (2000, 45), environmental factors were not considered to have any great significance with respect to the long-term safety of the repository; they were principally concerned with the implementation-related factors of investigations and the construction and operation of a repository.

In the latter part of 1985 TVO submitted their area selection research material to the authorities. The company presented a total of 102 rock blocks that were considered to be suitable for further research. Of these 101 were 'a result of the systematic selection and elimination process' (Vieno et al. 1992, 22). Based on a statement by the Ministry of the Environment, 17 areas were eliminated and 12 were redefined (mainly as the result of conservation area plans). The Radiation and Nuclear Safety Authority stated that the selection ought to pay particular attention to the geological variations of the areas. Furthermore, Olkiluoto in Eurajoki was chosen on the basis of a separate definition. According to the company's safety analysis, the site of the NPP was in a special position because the proximity of the facilities would reduce the transportation of SNF. The other reason given was the fact that because of the use of the rock block identification method, coastal areas were under-represented because the method used was generally thought to be unsuitable for the study of such sites (McEwen and Äikäs 2000, 9, 46). As far as coastal areas were concerned TVO's interest in selecting enough different research surroundings was met by the choice of Olkiluoto.

In 1985 some features of the voluntary approach were also adopted in Finland, when TVO approached possible candidate municipalities in writing. The company notified every one of the 66 municipalities which contained a potential research area. The importance of local acceptance increased, as the positive statement of the municipality to the locality of the siting was made a precondition of the new Nuclear Energy Act which was then under preparation and which became law in 1987. Furthermore there was in sufficient information in relation to the comparative long-term safety between the investigation sites. Based on the safety analysis conducted by TVO from 1985, according to which the suggested potentially suitable sites provided a possibility for safe final disposal, the emphasis of the siting programme was replaced by other qualifications such as public acceptance and land ownership (McEwen and Äikäs 2000, 9, 47).

However, site selection was not based solely on a voluntary principle. In the early part of 1986 Raumolin, the manager of TVO's nuclear waste office, emphasized that the local authorities in whose area a possible research site was situated had not been asked to give an opinion on the research. Thus, to quote Raumolin 'the fact that a local authority has not as yet said anything, does not disqualify the locality from further investigation' (*Finland's Municipalities* 1986).

In March 1986 TVO chose Kalliolampi area in Ikaalinen municipality as its first research area.[3] Very soon, however, the company withdrew from Ikaalinen because of local opposition. The Chernobyl disaster of April 1986 also had a substantial impact on the political atmosphere around the nuclear power issue. During 1986 the company held discussions about the project with local representatives based on contacts made by the local authorities after TVO's information letter. TVO discussed attitudes to the conduct of the research, even if it did not legally require permission except from the owner of the land (Posiva 1999a, Appendix 7, 11; viz. also *Finland's Municipalities* 1986). The aim was to clarify the municipalities' possible voluntary attitude, or even their willingness to commit themselves to a long-term area research programme (McEwen and Äikäs 2000, 9). These contacts also enabled TVO to begin to establish community relationships between the political hierarchy of the potential final disposal municipalities and the company.

In 1987 TVO began preliminary site characterizations for the selection of the site for the final disposal repository of SNF. TVO selected five research areas for preliminary disposal site studies. These were: Olkiluoto in Eurajoki municipality, Veitsivaara in Hyrynsalmi municipality, Kivetty in Konginkangas municipality, Romuvaara in Kuhmo municipality and Syyry in Sievi municipality. In April 1987 TVO's vice-chairman, Haapala, gave reasons for the selection of the five research localities from among the 85 research areas by stating that the municipalities were prepared to accept the research (*Helsingin Sanomat* 9 April 1987; viz. also Litmanen 1994, 72–3; McEwen and Äikäs 2000, 9–10). The research results were published in 1992. On the basis of these findings the company maintained that all of the investigation areas were positive in terms of their suitability from a final disposal point of view. Nevertheless, Hyrynsalmi and Sievi were excluded from detailed characterization because their bedrock structure was found to be more complicated than in the other areas, and it was not thought that further investigation would give sufficient information to eliminate the uncertainties (Posiva 1999a, App. 7, 12). Thus, detailed site characterization included only three sites: Olkiluoto in Eurajoki municipality, Kivetty in Konginkangas municipality and Romuvaara in Kuhmo municipality.

In a newspaper interview in 1995 TVO's principal geologist, Timo Äikäs, admitted that 'Kivetty in Äänekoski and Romuvaara in Kuhmo are included for competitive reasons with their railways. Concentration on technology and equipment would give one municipality too much influential advantage, for example in real estate taxation' (*Aamulehti* 14 April 1995). TVO also presented, albeit in stages, the evaluation principles of the siting process to the people in the municipalities under investigation. In November 1995 Äikäs stated that the research areas were similar with respect to their geological structures, so that the final site choice would be influenced strongly by the service capability of the locality as well as the cost of the excavation work (*Sisä-Suomen Lehti* 16 November 1995).

In the TILA-96 safety analysis commissioned by TVO, the importance of geology as a decisive siting criterion was discarded. The analysis (Posiva 1996) stated that no extraordinary characteristics that would deviate from the customary qualities of Finland's bedrock are required from a site in the crystalline bedrock to ensure the long-term safety of a repository for spent fuel. According to the analysis, the diversities between the properties and structure of the bedrock within the investigation area were of greater importance than the variations between different areas.

In 1997 another new site was selected. The nuclear waste management company Posiva, established by TVO and IVO, chose Loviisa, in the same way as Olkiluoto, because of its special position as the location of IVO's NPP (Posiva 1999b, 6). In 1997 at the start of the EIA process, Äikäs of Posiva concluded that the candidate municipalities 'are not placed in an order of preference based on any tables, if the bedrock of all of them is suitable. In the Finnish system we must be able to rely on the research results, and the decision must be made based on them. In the end the decision is political' (*Loviisan Sanomat* 31 January 1997). The preconditions of site selection were taken up in an EIA newsletter by Posiva (*Posiva's EIA leaflet* 2/97). The sentence in the leaflet 'the quality of the underground bedrock is decisive on a long-term basis' referred to geological selection criteria, when, by contrast, the municipality's own interest related to a flexible selection process. 'The above ground factors which have an effect on the area's suitability are among other things, technical aspects, environmental protection and land usage, social factors and the municipality's own willingness to have a final disposal repository' (*Posiva's EIA leaflet* 2/97). These five points were defined by examples. No mention was made of the emphasis on the criteria or who made the selection. Nevertheless, Posiva's information manager, Osmo Kurki, still presented the willingness of the municipalities as a conditional criterion in the latter part of 1998, when he said that a positive attitude in the municipalities could have an effect on the siting decision (Kojo 2002, 56–8). Posiva had clearly changed its interpretation in respect of the timing of the municipal decision making. Previously, Posiva had emphasized the formal decision making of the DiP procedure, in which the statements of the municipalities would be requested only after Posiva's DiP application, but now the company was eager to hear these statements in advance. The formal decision-making schedule was used as an argument against demands by the local movements to stop the site investigations.

Posiva mentioned the digression from the systematic selection process only indirectly in its environmental impact statement (EIS). In the EIS Posiva set out the main stages of the selection of the disposal site from 1983 onwards, stating briefly that a separate geological description was made of the suitability of Olkiluoto in Eurajoki for further research, 'because Olkiluoto as a site of a NPP had a special position in the area selection process' (Posiva 1999b, 5). By contrast, the nuclear waste company referred to the evaluation group of the International Atomic Energy Agency (IAEA). The group recommended that 'the objective in the siting is not "to find the best possible site" but the selection of a "suitable" site which meets the safety regulations of the final disposal based on a multi-barrier approach' (Posiva 1999b, 8). In spite of the safety objective stretching over thousands of years, the company did not deem it sensible to list the investigation sites in the order of preference based on geological safety criteria, even though geology is an essential factor in long-term safety (viz. Posiva 1999a; b).

In 1999 Posiva made a statement in its DiP application that in all four of the areas that had been researched it was possible

> to show sufficiently large and sufficiently integrated rock capacities, where the conditions are chemically and mechanically sufficiently suitable and stable to provide a sufficient barrier to prevent the release of radioactive substances, and which are suitable for the construction of final disposal facilities. (Posiva 1999a, App. 5, 28)

In other words, the best site was not chosen on a geological basis because the best geological site would have been politically impossible to select. Moreover, research and compensation costs would undoubtedly have been considerably higher. Posiva (1999a, App. 5, 35) also stated that the containment capacity of the final disposal facility would be effective without the influence of the bedrock and nature. The conclusion of the safety analysis was that 'no surveyed area can be regarded as clearly safer than the others, neither does the safety analysis give any reason to discard any of the alternatives' (Posiva 1999a, App. 5, 40). One clear difference could be seen in the quality of the ground waters, which were more saline at the coast than inland (Posiva 1999a, App. 5, 3; Posiva 1999a, App. 7, 158.) This did not, however, prevent Posiva from choosing Olkiluoto as the final disposal site. According to Posiva, 'some differences can be noticed in the rock fractures and flow conditions between the different sites, but the differences have no essential meaning from the perspective of long-term safety' (Posiva 1999a, App. 5, 40).

Political criteria thus became important factors, together with geology and the environment. The company stated that

> in evaluating the suitability of the site, attention must be paid not only to the geological conditions related to long-term safety but also to the aspects of the implementation of the final disposal. An essential factor regarding the implementation is also to gain local acceptance for the operation. (Posiva 1999a, App. 7, 15)

Posiva also changed its previous opinion on the progress of a strictly formal decision-making process, especially when speaking to the local decision makers. The company was hoping to receive clear positive signs from the municipal councils participating in the siting competition before the DiP process, whereas it had earlier stressed a procedure based on the formal DiP process and the timetable of the government (Kojo 2002, 57–8). This was a clear attempt to minimize the elements of uncertainty concerning local acceptance and thus to safeguard the approval of the DiP application.

Before the publication of the DiP application and the criteria in May 1999, the company had already made its own selection. According to Posiva, 'the site will be chosen based on long-term research in the way that it will offer a basis for safe final disposal with the greatest assurance' (Posiva 1999a, App. 7, 13).

Posiva did not present the final evaluation principles used to evaluate the suitability of the alternative sites until the DiP application, in other words after the site selection, which was felt to be late from the point of view of the municipalities that had been studied. The evaluation principles were focused on the following criteria: long-term safety, the suitability for construction, the expansion of final disposal space, the use of final disposal repository, social acceptability, land use and environmental impacts, infrastructure and costs (Posiva 1999a, App. 7, 13–26; also McEwen and Äikäs 2000, 178–89.) The government decision in relation to the safety of the final disposal, which defined general requirements, was made in March 1999, only two months before Posiva made its selection public. An awareness of the necessity to justify the DiP only became obvious at a considerably late stage (viz. McEwen and Äikäs 2000, 173–5).

Twenty years of arm-wrestling in Eurajoki

The question of nuclear waste was already under discussion during the early part of Teollisuuden Voima's NPP project, when in December 1973 the municipal council of Eurajoki approved the plan required by the licence to build an NPP. The prerequisite of the approval was that the SNF should not be buried in Eurajoki's bedrock. This view was based on the information provided by the nuclear power company. Local citizens took the promise seriously, and as a result over the years the general view that nuclear waste would not remain in Eurajoki persisted in the municipality.

The question of nuclear waste emerged again in 1977, when TVO proposed an extension of the plan of the NPP area to include the area of Kaalo in Olkiluoto. People who opposed the idea of depositing the nuclear waste in Eurajoki, insisted that the construction plan should include defined limitations in respect of the storage of radioactive waste, as limitations had not been registered anywhere. In January 1980 the local government of Eurajoki proposed by four votes to three (with one abstention) to consent to the municipal council for the extension of the construction plan. The council approved the proposal (with 14 against and 13 abstentions) with the addendum that the plan must not be submitted to the approval of the provincial government until TVO had given an undertaking that no final disposal of high active nuclear waste would take place in the area (Litmanen 1994, 98–101). TVO gave a written undertaking to this effect on 31 March 1980. The company underlined in its written undertaking that the final disposal would not depend solely on the company, and that it was not responsible for any other decisions that might be taken (TVO 1980). The opponents of final disposal subsequently recorded an objective in the Eurajoki municipal report that the municipality should ensure that the waste did not remain in Eurajoki.

Between the years 1973 and 1980 TVO negotiated with foreign companies on several occasions about a reprocessing agreement for SNF. When the TVO negotiations on the reprocessing agreement were concluded in the early months of 1980, it was clear that SNF would remain at the site for the time being, and that the company needed more storage space in Olkiluoto. TVO began the construction of a separate pool storage facility in Olkiluoto in 1984. The storage was taken into service in October 1987 (Björklund et al. 1994, 145–9).

In 1987 TVO chose Olkiluoto in Eurajoki as one of the five sites for preliminary characterization. As the site selection process for the final disposal of SNF was approaching its next elimination stage, in December 1992 councillors Juha Jaakkola (Centre Party), Paavo Majaneva (Left Alliance) and Antti Puosi (Centre Party) proposed that the sentence in the municipal report for 1993–97 which stated that 'the council endeavours to act so that no final disposal of high-active nuclear fuel will take place in Eurajoki municipality' should be reworded as follows: 'the council must act so, that no final disposal of high-active nuclear fuel will take place in Eurajoki municipality' (Kojo 2004a, 139). The more precise statement of the opponents' opinion was accepted, but it did not have the desired effect on TVO, as the company still selected Olkiluoto as one of its investigation sites. It must also be noted that even if the majority of Eurajoki's municipal council opposed final disposal in Eurajoki, the council had issued a statement in support of the application for a DiP on the construction of the new NPP unit in November 1991. The votes were split 17–10.

As the siting process progressed, the proponents of the project began a discussion in Eurajoki which aimed to remove the above statement from the municipal report. As a result of the municipal elections in October 1992 there was a change in the balance of power on the municipal council. In October 1993 the local party group of the National Coalition Party proposed that the municipality should immediately establish an organ for cooperation with TVO in order to ascertain the financial impacts of siting the final disposal repository on Eurajoki (National Coalition Party 1993). The local party group stated that disposal is a high-tech industry which would have no negative impact. Furthermore, the repository would ensure a flow of tax revenues to the municipality. In December 1993, however, following a vote in which the chairman, Juha Jaakkola, cast the decisive vote when the scores were tied at 13–13 (with one abstention), Eurajoki's municipal council approved the inclusion of the following sentence for the municipal report: 'The council must act in such a way that no final disposal of high-active nuclear fuel will take place in Eurajoki municipality.' Nevertheless, the arguments surrounding the municipality's strategic line continued. On 12December 1994 Eurajoki's local government approved the proposal by Keijo Kuusisto (National Coalition Party) that the sentence expressing a negative attitude to the final disposal should be removed from the municipal report. Markku Palonen (Social Democratic Party) spoke in favour of the proposal, stating that omitting the sentence forbidding the waste in the municipal report would undermine the interaction between the nuclear power company and the local authorities (*Satakunnan Kansa* 13 December 1994). According to Palonen, the removal of the sentence may well cause the company to take a positive attitude to the change in the law on real estate tax, which would increase the local revenue. Councillor Majaneva, seconded by Maritta Ristilä (Centre Party) and Jaakkola, proposed that the sentence should be added to the municipal report. The sentence including opposition to final disposal was finally removed from Eurajoki's municipal report following a vote of 15 to 10 in December 1994. As a result the stand of the municipality regarding the final disposal project was neutralized.

What was the origin of this sudden change of approach – which had occurred in the middle of the council's term? According to the newspapers (*Satakunnan Kansa* 13 and 14 December 1994), the Eurajoki councillors were pressurized in to changing their minds by a planning proposal. TVO was simultaneously applying for a change of the plan for the island of Kuusisenmaa, which was only some one hundred metres from Olkiluoto island, but is located in the territory of the town of Rauma. The councillors of Eurajoki were given to understand that the site of the final disposal facility could have been relocated to Rauma. This would also have meant that the real estate tax revenue (approx. €1.17 million) from the repository would have found its way into the coffers of the neighbouring municipality. TVO's head of information forbade the use of pressure. The company, however, dangled the carrot of the financial benefits of the final disposal project in front of the municipalities. In addition to the pressure exerted by the possible loss of real estate income, the Nuclear Energy Act had also been amended that year, as Parliament forbade the export and import of nuclear waste. Following this change in the law the co-operation of nuclear power companies in SNF management was realized through the establishment of Posiva in 1995. Although the municipal council of Eurajoki had opposed the retention of nuclear waste in the municipality, the opinion of some councillors had obviously changed in the sense that they now only opposed the final disposal of SNF in Eurajoki if the fuel had overseas origins. Furthermore, as stated earlier the composition of the municipal council had changed following the municipal elections of 1992.

Liaison groups as arenas for cooperation

Towards the end of the 1980s a liaison group, concentrating on bedrock research, was established between the municipality of Eurajoki and TVO (Posiva from 1996). This group also acted as a follow-up and steering group of the environmental impact assessment process during the period 1997–99. The institutionalization of the agreement between Eurajoki municipality and TVO can be seen in the cooperation agreement that was signed in August 1995. At the signing of the cooperation agreement, the municipal manager, Juhani Niinimäki, stated 'that the objective of the municipality was above all to maintain the financial stability regarding the tax revenue' (TVO 1995). Under the agreement the parties attempted to establish a permanent and stable format for reciprocal interaction. According to the agreement, the municipality would ensure that the decisions related to the activity of the company were made objectively, with the necessary expertise and without delay. In this way the company wanted to safeguard smooth local decision making in future. For its part TVO would ensure that the interests of the municipality were advanced and taken into account in the company's operations. One interesting element was also the principle of compensation agreed on. TVO undertook to compensate the municipality's proven costs that arose due to the operations of the company. However, compensation would be paid only when separately agreed (Kojo 2007a, 42; 2007b). The commitment of TVO to compensate could be interpreted as a positive sign of readiness on behalf of the company's board to negotiate on monetary benefits as part of the development of nuclear waste management.

According to the chairman of Eurajoki's municipal council, Juha Jaakkola, preparations for the shaping of the municipality's opinion regarding the DiP application began around 1996 (Jaakkola 2000). In February 1996 in the liaison group between Eurajoki municipality and TVO, Posiva's managing director, Veijo Ryhänen, stated that an early decision on the disposal site would be more advantageous both to the power company and to Posiva. The local authorities replied that Eurajoki wanted to be among the alternatives under consideration. The reply was unofficial and cautious, because the opinions of the councillors were clearly divided (Kojo 2004a, 142). Once again, as in TVO's commitment to the principle of compensation, the nuclear industry was active in initiating dialogue with the municipality about the siting of the repository.

The matter was taken up again in the liaison group in February 1997, when it discussed 'further development of the prerequisites for the cooperation between the municipality and TVO' (The Municipality of Eurajoki

1997a). The concrete project was the final disposal plan. In April 1997 Posiva announced that in the EIA process the municipalities' visions would also be taken into account (The Municipality of Eurajoki 1997b). One of the effects of Posiva's active quest for visions was that the formulation of the municipal strategy was initiated in Eurajoki. The formulation of a strategy was influenced not only by a discussion of the selection of the final disposal site, but also by the weakened financial situation of Eurajoki municipality and the desire to safeguard its tax revenues. During the period 1984–90 TVO had paid local presumptive tax to the tune of €18.5 million, which was about one-third of Eurajoki's annual tax revenue. The practice of the levying the presumptive tax on business income was abolished from 1991 onwards, which caused Eurajoki municipality a loss of more than €2.35 million in tax revenues in 1991 and 1992. The municipality was, however, compensated by the state of Finland for the loss of income (Vahekoski 1992, 85–6; Björklund et al. 1994, 190). The presumptive tax was replaced by a real estate tax based on the value of property from 1993 onwards. The setting of the real estate tax was not straightforward as such, not least because of the annual age reduction of the NPP buildings.

The opening of the dialogue by the nuclear energy industry about the future vision of the municipality did produce some results. The preparations led to the discussion of a project for a new old people's home to replace the Vuojoki Manor old people's home. Municipal manager Niinimäki introduced the idea in the liaison group in August 1997 (The Municipality of Eurajoki 1997c). On the same occasion reference was made to the project of the municipality's multi-purpose hall which was at the planning stage at this time. In October 1997 Eurajoki's local council decided that the municipal manager and the planning secretary should make a proposal for the programme and schedule of a strategy project. The task laid down by the municipal council was to create a long-term vision for the future of the Eurajoki municipality, with the deadline for the completion of the vision being set as the end of April 1998.

In order to establish the drafting of the municipality's strategy an extended working group including representatives from the Eurajoki municipality and the local business community was established which met for the first time in January 1998. In the extended liaison group councillors Altti Lucander (National Coalition Party) and Markku Palonen (Social Democratic Party) as well as the municipal manager Niinimäki concentrated on discussing the opportunities and threats presented by Olkiluoto to Eurajoki's municipal strategy. The threat they perceived as resulting from Olkiluoto was the fact that nuclear waste would remain above ground in Eurajoki, or that it would be transported to another locality, and that the municipality would lose the tax revenue in either case. The competition between municipalities and the fear of loss of tax revenues featured prominently in the preparation of the strategy. In 1997 the selection of Loviisa as a new candidate site intensified the competition between the municipalities. Lucander, Palonen and Niinimäki perceived another threat in the government's action. If the government was to make the decision about the siting of repository for SNF, Eurajoki would not have a leg to stand on when negotiating with the nuclear energy industry. Should Eurajoki block the final disposal with its right of veto, the three of them considered it possible that the government could make the decision about the disposal of the waste. This would mean that the waste would stay in the municipality, but possibly that they would not receive any compensation. This threat could only have arisen if no other municipality had approved the final disposal project. One aspect of the final disposal was the possibility that the municipality could negotiate an additional financial benefit to be paid as compensation until the real estate revenue took effect. As an example of compensation for the early years of 'low income', reference was made to the old people's home project (The Municipality of Eurajoki 1998a).

The progress of the final disposal project was also seen as helpful for the establishment of the third NPP unit in Olkiluoto. The real estate tax of the new unit, anticipated to amount annually to around €2.5–3.4 million, was perceived by Lucander, Niinimäki and Palonen as a way of safeguarding the municipality's permanent autonomy, since the ending of nuclear energy production in Olkiluoto would decimate its tax revenues and endanger the municipality's autonomy. In conjuring up these threatening images Lucander, Niinimäki and Palonen did not consider the possible environmental or image risks that might be caused by the final disposal repository. On the contrary, it was feared that a refusal to invest millions would impair the image of the municipality on a national level. There were also other fears regarding the discontinuation of electricity production in Olkiluoto. So much so that it was feared that Eurajoki could lose its 'relative advantage' as plants replacing nuclear energy do not meet with corresponding political opposition, and they could therefore be built elsewhere. In other words, they wanted to retain Eurajoki's position as an oasis of the nuclear energy industry. In the nightmare scenario the municipality's financial status would lose its existing strength if no further construction work took place in Olkiluoto, and the activity would slowly cease (The Municipality of Eurajoki 1998a). The future of Eurajoki was thus very firmly linked to that of the nuclear energy industry in Olkiluoto.

At the same time, in January 1998, as the extended working group was preparing the municipal strategy, the Vuojoki working party, which consisted of the representatives of Eurajoki municipality, Teollisuuden Voima and Posiva, began its work. The initial task of the working party was to draft the details of the above-mentioned old people's home project and the utilization of the Vuojoki Manor. Later the working party was assigned further preparatory tasks (for more on the negotiations, see Kojo 2008). The establishment of the Vuojoki working party followed the existing political culture, characterized by the harmonization through negotiation of the interests of the nuclear energy industry and the municipality.

Some of the local politicians can be seen as a driving force for the local development concerning the municipality's attitude to nuclear waste disposal. In 1993 the local party group of the National Coalition Party (1993) had already proposed that the municipality should approve the siting of the repository in Olkiluoto. Councillors Altti Lucander and Markku Palonen were also active in 1996. Palonen (*Satakunnan Kansa* 30 January 1996) considered the possibility of using the Vuojoki Manor as the headquarters of the power companies and replacing the old people's home with a new one. Lucander (*Länsi-Suomi* 7 July 1996) emphasized that the municipality should take a stand on the siting issue. In March 1998 local Social Democrats stated in an undertaking that the nuclear communities have moral responsibility for final disposal of SNF (*Satakunnan Kansa* 29 March 1998). In June 1998 – during the preparation of the above-mentioned municipal strategy – local Social Democrats demanded that the municipality should decide as soon as possible whether or not it would approve the siting of the repository in Olkiluoto (*Satakunnan Kansa* 5 June 1998). According to the local Social Democrats a positive attitude would assist in the development of the basic health and social services of the municipality, with a new facility for health and social care mentioned as an example (Kojo 2007a, 42–3). As a result, the siting issue had strong proponents among the local politicians in Eurajoki.

One result of the work of the extended liaison group of Eurajoki municipality was the presentation of the municipal strategy delivered to the council in September 1998. In December 1998 Eurajoki's council approved the Olkiluoto vision by 20 votes to 7 as a part of the municipality's strategy. The vision contained a positive attitude to the further construction of nuclear energy and a final disposal repository for SNF (The Municipality of Eurajoki 1998b, 10–11). Thus the attitude of the municipality had changed completely in only four years, given that the sentence forbidding the disposal of SNF in Eurajoki had only been voted out of the municipal report in December 1994. Evidently such a rapid change would not have been possible without the efforts of some local politicians as proponents in local decision making.

Compensation agreements as a result of cooperation

Because of the financial situation of the Eurajoki municipality, and in order to safeguard its real estate tax revenue, in 1988 it proposed to the Ministry of Finance that the real estate tax be raised for NPP buildings from 1.8 per cent to 2.5 per cent. The municipality's suggestion for a tax increase did not go down well with TVO. It recommended instead that the municipality promote projects that would lead to a widening of the tax base.

A change in the taxation as well as the use of Vuojoki Manor was discussed during a visit to Eurajoki in May 1998 by the then Minister of Education, Olli-Pekka Heinonen (National Coalition Party). The minister announced that the government had discussed the suggestion, and that the final disposal municipality could be granted considerable real estate tax revenues before the repository was operational. Such an idea was included in the Energy Policy Report of 1997 (Council of State 1997, 46–7). In that way the government was also considering offering 'a carrot' to the future final disposal municipality.

During August and September 1998 leading local politicians discussed the possible forms of compensation for the repository. The aim was to trade with the Vuojoki Manor, which would be the means for compensation in the form of a new old people's home. In addition, compensation would be sought for the planning phase of the facility between 2000 and 2010, as the real estate tax would not be paid during that period. The core group of local politicians decided that the negotiations with TVO would start after they had obtained an estimate of the costs of a new old people's home. They also planned to meet the then Minister of Trade and Industry, Antti Kalliomäki (Social Democratic Party), and the Minister of Finance Jouko Skinnari (Social Democratic Party) to discuss the possibility of receiving some real estate tax payments before the operation phase of the final disposal facility (Kojo 2007a, 43; 2008). Taken as a whole, due to the activities of the nuclear industry in initiating discussion on the municipal strategies and the activity of the proponents inside the local decision-making system, local dialogue resulted partly in a negotiation process which prepared the way for compensation.

Following a government decision, from the beginning of 1999 the property tax on NPP buildings rose to 2.2 per cent, which raised the industry's annual tax liabilities by some €23.5 million. The mounting expense and the duties of the TVO board caused the company to appeal to the Administrative Court against the real estate taxes, initially for 1993, and later, in December 1999, for 1994. The appeals provided an additional boost to the compensation negotiations, as the appeals were a further shock to the foundations of Eurajoki's local economy.

At the beginning of March 1999, a proposal was made to the TVO and the municipality liaison group for alternative compensation options for the payment of real estate taxes. These included, for example, Eurajoki's ice-stadium project, a development fund for business projects in the municipality, the planning agreement of the Olkiluoto areas, plus various loans as well as the sale or leasing of the municipality's water areas. The representatives of the municipality proposed that the preparation of taxation matters be handed over in its entirety to the Vuojoki working party (Kojo 2008). Before this time TVO's appeal against the real estate tax of 1993 had already been discussed in an unofficial working party between the municipality and the company.[4] The other economic negotiations on agreement relating to the final disposal project were also handed over to the Vuojoki working party for preparation.[5]

The Vuojoki working party thus became a more central preparation organ for decision making, where the nuclear energy industry could have a direct effect on the decision making relating to its own interests within the local democracy.

At the end of March 1999 the local government of the Eurajoki municipality presented the ministry with a written proposal for the final disposal project. The municipality had drafted for the government a wish list with three points. Firstly, the municipality proposed that the legislation on state subsidies should be changed in such a way that the real estate tax of a NPP unit and a repository were excluded from the calculatory tax base when accounting the tax revenue equalization of the municipality. In 1998 the amount of this equalization was estimated to reduce the municipal revenue by €0.72 million and in 1999 €0.67 million. The municipality tried to keep the whole amount of real estate tax paid by the nuclear industry in its own coffers.

Secondly, the municipality proposed that the real estate tax percentage of a repository for SNF should be considerably higher than the percentage for a NPP building and that the percentage should be included in the real estate tax legislation. The municipality stated that the siting of a new and globally unique facility would incur expenses for the municipality (Kojo 2004a, 148). Originally, the municipality argued from the point of view of safety issues, but it changed these arguments after the critical feedback on the draft of the proposition by the nuclear industry in the Vuojoki working party, where the representatives of the nuclear energy industry criticized Eurajoki's proposal for taxation change (Kojo 2007a, 44; 2007b; 2008). Posiva's managing director Ryhänen stated that Fortum did not seem to have any sympathy for the project, which would mean a rise in the property tax level in spite of the planned agreements. For his part, TVO's representative, Ami Rastas, stressed that 'the final disposal must be absolutely safe, and there must be no connection whatsoever between money and the prospect of safety' (The Municipality of Eurajoki 1999). TVO also refused to cancel the appeal against the 1993 real estate tax, but offered the local authority advantageous loan arrangements and their participation in the ice-stadium project, as a form of lifeline (Kojo 2008).

Thirdly, the municipality wanted the compensation to be paid immediately after the site selection whereas the real estate tax of the repository was to be paid in full by Posiva only after the completion of the repository (according to the schedule, in 2020). Thus the municipality wanted to be paid a substantial financial compensation either as real estate tax or in the form of some other kind of payment immediately after site selection and at the beginning of the research phase (Kojo 2004a, 149).

The report by the Vuojoki working party on the possibilities for the use of the Vuojoki Manor was ready by 15 April 1999. The working party also drafted the propositions for a land rental and loan agreement between Eurajoki municipality and Posiva. In this way it used the preparation power granted in the liaison group between the municipality and TVO. In order to fine-tune the agreements a separate small working party was established. The report was discussed at the municipal council meeting on 26 April. This stated that as the old people's home projects of the Eurajoki municipality would require development and expansion in the near future, the municipality had been investigating whether or not the Vuojoki Manor could be replaced with a new, modern old people's home, and at the same time propose an alternative use for Vuojoki. It was in Posiva's interest to consider Eurajoki as a final disposal site for SNF, in which case the Vuojoki Manor could be developed as office space for the company alongside its more general use. The local government suggested that the council approve the agreement on condition that Posiva should consider only Eurajoki as the final disposal site for SNF. Another part of the agreement package was TVO's funding for the ice-stadium project.

The chair of the local government and the deputy chair of the Vuojoki working party, Matti Valtonen (Social Democratic Party), explained the newly approved agreement by stating that the municipality had informed Posiva that they really wanted to site the repository. Furthermore, according to Valtonen, applying for the DiP in Loviisa as well would mean restarting the negotiations. Posiva's managing director Ryhänen stated that the company could still consider how many municipalities would be included in the application, and stressed the positive interaction with the municipalities proposed for final disposal. According to the managing director,

> this Eurajoki project is quite unique. It does not, however, indicate that the final disposal repository will automatically be in Eurajoki. It does, however, prove that the municipality is interested in the matter, and the project had already appeared in the municipality's strategy as an interesting project. (*Länsi-Suomi* 24 April 1999)

On 3 May 1999 Eurajoki's municipal council approved the Vuojoki Agreement by 20 votes to 7. Under the agreement Eurajoki municipality leased the Vuojoenlinna estate to Posiva, which the municipality owned and whose early nineteenth-century manor had been an old people's home. Furthermore, Posiva loaned the municipality some €6.89 million (€6.39 million + €0.5 million) for the construction of a new old people's home. For its part the municipality was obliged to pay the instalments and interest on the loan with the rent it would receive from the company. The agreement stipulated that the manor would be renovated for its new use.[6] As a part of the compensation TVO paid a shareholder loan of €0.5 million to the Eurajoki ice-stadium company for the construction of a new stadium. This shareholder loan from TVO and the €0.5 million part of Posiva's loan (in total €1.0 million) compensated the municipality for TVO's 1993 real estate tax (Kojo 2008).

On 26 May 1999 Posiva submitted the DiP application for the final repository for SNF to the government with the Eurajoki municipality as the only site. According to Posiva's managing director Ryhänen (*Satakunnan Kansa* 27 May 1999), Eurajoki's initiative and clear desire to secure the final repository decided the matter. Other grounds for the selection were that Olkiluoto already had most of the spent fuel, so that with the repository in Olkiluoto a minimum amount of transport would be needed, with connections by both land and sea. In addition, Olkiluoto already had important equipment for the operation of nuclear stations, and Olkiluoto offered better facilities than Loviisa for the expansion of the repository, should the amount of waste increase.

Even if the Eurajoki municipality was promised compensation for the repository, the municipality was still experiencing considerable economic difficulties. An extra expense of €1.18 million was the repayment of the real estate tax for 1994, resulting from the appeal by TVO in December 1999. The municipality demanded that the reimbursement be compensated in full by TVO and that the matter should be negotiated before the meeting of the municipal council on 24 January 2000. The statement of the municipality on Posiva's DiP application was on the agenda at that meeting. In the Vuojoki working party the municipality carried out negotiations with TVO on compensation for the reimbursement of the real estate tax. TVO offered to buy the municipality's share of the Olkiluoto and Orjasaari water area for €0.84 million and to arrange funding for the Business Development Fund in Eurajoki to the sum of €0.25 million. Moreover, TVO offered a very cheap loan of €2.35 million, at an interest rate of 0.75 × 3 months or 6 months Euribor to help the municipality with its liquidity problems for paying the real estate tax reimbursements of 1993 and 1994 (Kojo 2007a, 45–6; 2007b, 2008).

At its meeting on 19 January 2000 the local government stated that the starting point of the municipality's statement is that only SNF from Finland's NPP can be disposed of in Olkiluoto in accordance with the Nuclear Energy Act. The municipal council of Eurajoki discussed the statement on the application for a DiP on 24 January 2000. Councillor Paavo Majaneva (Left Alliance) pointed out that during Olkiluoto's construction stage the impression was given that the nuclear waste would be transported abroad. Majaneva made a counter-proposal according to which the final disposal of the nuclear waste should be postponed. Raimo Järvi (Left Alliance) seconded the proposal. The municipal council, however, made a positive response by 20 votes to 7.[7] At the same meeting the council also approved the deal concerning the sale of the municipality's Olkiluoto and Orjasaari water areas to TVO, the loan agreement with TVO, and the agreement for the establishment of a business development fund in Eurajoki. TVO undertook to provide the fund with some €50,000 per year for five years.[8] The proponents of the project defended the decision by stating that it was the moral responsibility of the municipality as it benefited from the NPPs, and that this safety was confirmed by the Radiation and Nuclear Safety Authority (STUK). The decisive local step towards the siting of the repository for SNF into Olkiluoto had therefore been taken.

Conclusions

The aim of this chapter was to analyse the site selection process for an SNF repository in Finland. The main question was why the Posiva nuclear waste management company chose the municipality of Eurajoki as a site for the repository. The question was approached from the viewpoint of the site selection strategy proposed by Sundqvist (2002).

The chapter was divided into three subtasks in order to help us to understand the context of the Finnish site selection process. The first subtask was to decipher the formation of Finnish nuclear waste policy. It was suggested that although the nuclear power company TVO had studied direct final disposal as an alternative since the early 1980s, it was only in the mid-1990s that direct final disposal became a joint aim of the two nuclear power companies operating in Finland. Furthermore, the reprocessing alternative was excluded following the passage of the amended Nuclear Energy Act. In practice, in the late 1970s TVO had already rejected reprocessing as too expensive. However, IVO's SNF management was based on reprocessing up until December 1996, when the last shipment of SNF was transported to Russia. Thus, IVO (which has been known as Fortum Power and Heat since 1999) was forced to reconsider its SNF management. The establishment of the joint nuclear waste management company, Posiva, in 1995, helped the nuclear power companies to focus their resources on the joint goal of direct final disposal.

The second subtask was to provide an overview of the changes in the site selection process that have taken place since the early 1980s. The chapter demonstrates, as Litmanen (1994) and Anttila (1995) have stated earlier, that geology as a site selection principle was given less emphasis after the early 1990s. Geological arguments for site selection were in evidence in early research reports – as occurred, for example, in 1992 when two out of five sites were not selected for detailed characterization because the structure of the bedrock was more complex than in the other sites. Other empirical data indicate that features stressing pragmatism in siting became stronger during the 1990s. In 1996 Posiva's safety analysis stated that safeguarding long-term safety does not require anything special from the bedrock of the site. Posiva emphasized in the DiP application of 1999 that the criteria applied were geological conditions related to long-term safety and the aspects of the implementation of the final disposal with the latter, including local acceptance. The function of geological criteria was only to ensure that the bedrock of the sites was good enough, thus the decisive criteria were actually non-geological in nature.

It is obvious that Posiva did not want to end up in a situation in which the municipality of Eurajoki as a nuclear community with a relatively high level of social acceptance among local residents could dictate the preconditions for siting. Thus, the other Finnish nuclear community, Loviisa, the location of Fortum's NPP, was chosen as an alternative candidate site to compete with Eurajoki in 1997. The inland sites in the municipalities of Kuhmo and Äänekoski were geologically suitable, but local opposition made them politically more unpredictable and hence unsuitable.

The third subtask concerned local decision making in Eurajoki, and in particular the negotiations on compensation levels that were held between the Municipality of Eurajoki and TVO and, later, Posiva. The loss of tax revenue was one important reason for the interest of some Eurajoki councillors in siting the repository in Olkiluoto. On the other hand the industry's problem was to safeguard a site for the repository. For TVO, Olkiluoto was a potential final disposal site, due, for example, to its limited need for transport and the existing infrastructure. The company used the economic benefits of the project as its trump card. The attitude of Eurajoki municipality to the siting of the repository became positive following the adaption of the Olkiluoto vision in December 1998, whereas five years earlier the municipal council had been prepared to act and prevent final disposal. The future vision presented by the municipal council now matched the interests of TVO. First the proponents of the project succeeded in removing the sentence opposing the final disposal from the municipal report, and after that they began working towards a positive response on the part of the municipality. As had already been the case at the beginning of the 1990s, towards the end of the 1990s the proponents were focusing not only on the siting of the repository in Eurajoki, but also the possibility of locating a new NPP unit in Olkiluoto, and steering its tax revenue into the coffers of the municipality. On 16 October 2003 TVO announced that the new 1,600 MW NPP unit would be built in Olkiluoto, with Loviisa being named as an alternative location.

A sufficiently large and sufficiently long-term compensation was put forward by the leading local politicians as a precondition for the approval of the siting of the repository. TVO and Posiva regarded it as ideal that one of the chosen candidate municipalities had volunteered. Posiva discussed the environmental impacts of the repository in public (see, for example, Hokkanen 2007), and at the same time Posiva and its main owner TVO negotiated with the representatives of Eurajoki on the political approval of the repository and its associated compensations. Posiva implemented the EIA process by stressing transparency and citizen participation, whereas negotiations on compensations between the contracting parties took place behind closed doors.

The three dimensions of the siting strategy as analysed – high-level nuclear waste policy, site selection and local negotiations – indicate that the year 1994 appears to have been a critical point in the Finnish site selection process. At the policy level transportation and reprocessing abroad was excluded as alternatives by the amended Nuclear Energy Act, which, in turn, had an impact at the local level, as the municipal council of Eurajoki was ready to neutralize its statement regarding the siting following the prohibition on the import of foreign nuclear waste introduced under the Act. The neutralization of the municipality's earlier (1994) negative statement on siting helped to begin closer discussions between the municipality and TVO about their mutual interests. In the steadily progressing site selection process the number of candidate sites was reduced. However, at the same time the competition between the municipalities was maintained, and thus the municipalities were forced to consider their visions regarding the siting of the repository.

The Finnish site selection strategy is characterized by pragmatism, where the criteria of geology and local acceptance are intermingled. The role of geology in the selection of Olkiluoto as a final disposal site was ultimately of only minor importance. The objective was not to select the best possible site. Neither was it possible to point out essential differences in the geology of the research sites regarding long-term safety. The siting was based on political expediency and competition between the candidate municipalities. Eurajoki as a Finnish nuclear oasis was preferred by the nuclear energy industry because of its existing infrastructure, the minimal need for high-level nuclear waste transportation and the permissive political culture which ensures that the interests of the nuclear industry are taken into account on a local level.

Acknowledgement

This chapter is based on the author's unpublished licentiate's thesis in political science *Building Acceptability. The Struggle Over Site Selection for the Final Disposal Repository from the Angle of Political Style.* (http://tutkielmat.uta.fi/pdf/lisuri00023.pdf) and the CARL Country Report *'Finland'*. The Country Report was presented in the CARL Workshop in Antwerp, Belgium, 30 November 2005. The author wishes to express his gratitude to the CARL Project for funding the translation costs of the Country Report. The licentiate thesis was written as a part of the Research Programme for Advanced Technology Policy ProACT (http://proact.ktm.fi) funded by the National Technology Agency.

This chapter was accomplished within the two following projects: Arenas for Risk Governance (ARGONA; http://www.argonaproject.eu/) co-funded by the European Commission (EC) as part of the sixth Euratom research and training Framework Programme (FP6) on nuclear energy (2002–2006). Contract no.: FP6-036413, and the research programme 'Power and Society in Finland' (http://www.aka.fi/valta) funded by the Academy of Finland.

A concise paper (Kojo 2007b; copyright the American Society of Mechanical Engineers, ASME) on this issue was presented and published in the 11th International Conference on Environmental Remediation and Radioactive Waste Management, September 2–6, 2007, Bruges, Belgium (paper number 7094).

The conclusions presented in the chapter are solely those of the author.

Notes

1. Björklund, Westerholm and von Bonsdorff (1994, 149) maintain that Wolter Westerholm took an obvious risk when he waived the nuclear waste management based on reprocessing, which followed the requirements presented by the MTI and the practice used in other countries. The management of TVO obviously knew of the dispute in Sweden regarding the charging of the Ringhals 3 reactor (viz. Sundqvist 2002, 78–105).
2. Interestingly, in an interview the Director General of STUK, Jukka Laaksonen, stated that he did not consider the direct final disposal of SNF to be a long-term solution. Laaksonen considered that the recycling of SNF would be preferred in the forthcoming decades because of ecological and economical arguments. According to Laaksonen, the repository would be needed for disposal of reprocessed vitrified high-level waste (*Loviisan Sanomat* 15 February 2008).
3. As part of the research and development programme, in 1984–85 TVO carried out in Katosuo, municipality of Lavia, some test hole drilling of the bedrock.
4. The minutes of the Vuojoki Working Party do not show who belonged to this unofficial working party, how often it met or what matters were discussed.
5. The new task of preparing real estate tax compensation was assigned to the Vuojoki working party by the liaison group of the municipality and TVO on 1 March 1999. According to the minutes of the liaison group, the municipality took the initiative. However, the contracting parties had agreed earlier, on 26 February, on the assignment in the meeting to the Vuojoki working party (Kojo 2008). The example shows that power was not in the hands of the liaison group, let alone the local council of Eurajoki.
6. The renovation of the main building of the Vuojoki manor was completed in 2005.
7. An appeal against the local council's decision was made to the Court of Appeal and the Supreme Administrative Court. This appeal was dismissed on 18 November 2000.
8. In 2004 TVO and Posiva decided to pay a further €150,000 during the period 2005–09 for the business development fund in Eurajoki (*Satakunnan Kansa* 19 August 2004).

References

Anttila, P. (1995) *Käytetyn polttoaineen loppusijoitus – loppusijoituspaikan valintaperusteet* (Final Disposal of Spent Nuclear Fuel – Basis for Site Selection). May 1995. Report YJT-95-07. Helsinki: IVO International Oy.

Björklund, N.G., W. Westerholm and M. von Bondsdorff (1994) *Ydinsähköä. Teollisuuden Voima Oy 1969–1994* (Nuclear Electricity. Teollisuuden Voima Ltd 1969–1994). Helsinki: Teollisuuden Voima.

The Council of State (1983) *Valtioneuvoston tekemä periaatepäätös ydinjätehuollon tutkimus-, selvitys- ja suunnittelutyön tavoitteista* (The Decision-in-Principle by the Council of State on 10 November 1983 concerning research, account and planning goals of the nuclear waste management).

The Council of State (1997) *Suomen energiastrategia. Valtioneuvoston energiapoliittinen selonteko 3.6.1997. VNS 5/1997 vp* (Finland's Energy Strategy. The Energy Policy Report of the Council of State. 3 June 1997. VNS 5/1997 vp). Reports of the Ministry of Trade and Industry 5/1997. Helsinki: Ministry of Trade and Industry.

Finland's Municipalities (1986) Municipalities have been interested in nuclear waste. Number 4, 18 February.

Haapala, E. (1988) Kymmenen vuotias YJT (Ten years old YJT). In *Ydinjätetutkimusta yhteistyönä* (Nuclear waste research as co-operation). The Nuclear Waste Commission of Finnish Power Companies YJT. 7–11. Helsinki: Imatran Voima, Teollisuuden Voima.

Hokkanen, P. (2007) *Kansalaisosallistuminen ympäristövaikutusten arviointimenettelyssä* (Public Participation in Environmental Impact Assessment). Acta Universitatis Tamperensis 1285. Tampere: Tampere University Press.

Jaakkola, J. (2000) Greetings from Eurajoki. Presentation in the seminar 'Current Nuclear Waste Problem'. Parliament 29 March.

Käytetyn ydinpolttoaineen loppusijoitus Suomen kallioperään (Final Disposal of Spent Nuclear Fuel into the Finnish Bedrock) (1982): Teollisuuden Voima Oy. Report YJT-82-46. The Nuclear Waste Committee of Power Companies [Helsinki].

Kojo, M. (2008) Compensation as Means for Local Acceptance. The Case of the Final Disposal of Spent Nuclear Fuel in Eurajoki, Finland. Paper presented in *the Waste Management 2008 Conference*, Phoenix, Arizona, 24–8 February.

Kojo, M. (2007a) Kompensaatiot teknologiahankkeen tientasoittajana. (Compensations as a paver of the way in a technology plan). In *Tiedepolitiikka* 32(2), 37–48.

Kojo, M. (2007b) Local Negotiations on Compensation. Siting of the Spent Nuclear Fuel Repository in Finland. In *Proceedings of the 11th International Conference on Environmental Remediation and Radioactive Waste Management* ICEM 07. Bruges, Belgium, 2–6 September.

Kojo, M. (2004a) Eurajoki – ei paras mahdollinen, mutta poliittisesti sopiva. (Eurajoki – Not the Best Possible, but Politically Suitable). In M. Kojo (ed.), *Ydinvoima, valta ja vastarinta* (Nuclear Power, Power, and Resistance). Helsinki: Like, pp. 127–58.

Kojo, M. (2004b) Yhteenveto (Summary). In M. Kojo (ed.), *Ydinvoima, valta ja vastarinta* (Nuclear Power, Power, and Resistance). Helsinki: Like, pp. 231–58.

Kojo, M. (2002) Lahjomattomien haukansilmien valvonnassa. Ydinjätteen loppusijoitushankkeen rakentaminen Posiva Oy:n tiedotusmateriaalissa (Under the Control of Incorruptible Eyes. The Acceptance Building of Nuclear Waste Disposal Plan in the Information Leaflets of Posiva Ltd). In P. Raittila, P. Hokkanen, M. Kojo and T. Litmanen (eds), *Ydinjäteihme suomalaisittain* (Finnish nuclear waste miracle). Tampere: Tampere University Press, pp. 36–66.

Lammi, H. (2004) Tarinat kovasta ytimestä (The Stories Concerning the Hard Nucleus). In M. Kojo (ed.), *Ydinvoima, valta ja vastarinta* (Nuclear Power, Power, and Resistance). Helsinki: Like, pp. 11–50.

Litmanen, T. (1994) *In the Bowels of the Bedrock, Hundreds of Meters Deep: Local Nuclear Waste Conflicts in Finland*. Unpublished Licentiate thesis in Sociology. University of Jyväskylä.

Loviisan Sanomat 31 January 1997; 15 February 2008. A local newspaper.

Länsi-Suomi 7 July 1996; 24 April 1999. A local newspaper.

McEwen, T. and T. Äikäs (2000) *The Site Selection Process for a Spent Nuclear Fuel Repository in Finland – Summary Report.* Posiva 2000–15. Helsinki: Posiva Oy.

Ministry of Trade and Industry (MTI) (2005) Nuclear Energy in Finland. Ministry of Trade and Industry. Energy Department.

Ministry of Trade and Industry (MTI) (1988) *Ydinjätehuollon kehittämistarpeet. Kauppa- ja teollisuusministeriön asettaman työryhmän selonteko* (The Development Needs of Nuclear Waste Management). Energy Department. Series C:18. Helsinki: Ministry of Trade and Industry.

Municipality of Eurajoki (1997a) The Liaison Group Between the Municipality of Eurajoki and Teollisuuden Voima. Meeting Number 1/97. The minutes on 24 February 1997.

Municipality of Eurajoki (1997b) The Liaison Group Between the Municipality of Eurajoki and Posiva in Issues Related to Bedrock Investigations. Meeting number 1/97. The memo on 24 April 1997.

Municipality of Eurajoki (1997c) The Liaison Group Between the Municipality of Eurajoki and Teollisuuden Voima. Meeting Number 3/97. Minutes on 25 August 1997.

Municipality of Eurajoki (1998a) The Extended Co-operation Working Party of the Municipality of Eurajoki and the Business Community. Meeting of the Co-operation Working Party. Memo 2/98. 24 February.

Municipality of Eurajoki (1998b) Eurajoki. Budget Draft of 1999. Budget Plan 1999–2001. Approved by the Local Council of Eurajoki on 7 December.

Municipality of Eurajoki (1999) The Meeting of the Vuojoki Working Party. Minutes on 23 April 1999.

National Coalition Party (1993) The Bill of the National Coalition Party Group of Eurajoki on the Establishment of a Liaison Group for the Local Council of Eurajoki Municipality (section 84). 18 October.

Posiva (1999a) *Käytetyn ydinpolttoaineen loppusijoituslaitoksen periaatepäätöshakemus.* (The Application of Decision in Principle of the facility for spent nuclear fuel). Helsinki: Posiva Oy.

Posiva (1999b) *The Final Disposal for Spent Nuclear Fuel: Environmental Impact Assessment Report.* Helsinki: Posiva Oy.

Posiva (1997) Posivan YVA-tiedote (*Posiva's EIA Leaflet*) 2/97.

Posiva (1996) Interim Report on Safety Assessment of Spent Fuel Disposal. TILA-96. Timo Vieno and Henrik Nordman VTT Energy. Posiva-96-17. December. Helsinki: Posiva.

Rasilainen, K. and S. Vuori (1999) *Käytetyn ydinpolttoaineen huolto. Suomalaisen suunnitelman pääpiirteet* (The Management of Spent Nuclear Fuel. The Main Outlines of the Finnish Programme). VTT tiedotteita 1953. Espoo: The Technical Research Center.

Raumolin, H. (1982) *TVO:n ydinjätehuollon toimintaohjelma ja aikataulut* (Programme and Schedules for TVO's Nuclear Waste Management). The Nuclear Waste Commission of Finnish Power Companies. Report YJT-82-55. Helsinki: Teollisuuden Voima Oy.

Richardson, P.J. (1998) *Raportti Suomen ydinjätesijoitukseen liittyvistä ongelmista* (The Report Concerning the Problems Related to the Final Disposal of Nuclear Waste in Finland). Ed. I. Simes. Copies of the Finnish Association for Nature Conservation 4/98. Helsinki: The Finnish Association for Nature Conservation.

Ryhänen, V. (1979) *Käytetyn polttoaineen huoltovaihtoehdot ja ulkomaiset palvelut.* (Alternatives and Foreign Services for Spent Fuel Management). December. The Nuclear Waste Commission of Finnish Power Companies. Espoo: Teollisuuden Voima Oy.
Ryhänen, V. (1985) Ydinjäte kallioon. (Nuclear Waste into Bedrock). An article in *Suomen kunnat* (Finnish Municipalities) 17, 30–2.
Ryhänen, V., J.-P. Salo, I. Mikkola and H. Raumolin (1981) *TVO:n käytetyn ydinpolttoaineen huollon vaihtoehdot.* (TVO's Spent Fuel Management Alternatives). September 1981. Report YJT-81–15. The Nuclear Waste Commission of Finnish Power Companies. Teollisuuden Voima Oy.
Sandberg, J. (1999) Päättikö eduskunta geologisesta loppusijoituksesta jo vuonna 1994? Käytetyn ydinpolttoaineen huoltoa koskeva lainsäädäntö ja viranomaispäätökset. (Did Parliament Decide on Geological Final Disposal as Early as 1994? Legislation and Decisions of the Authority Concerning the Management of Spent Nuclear Fuel). In T. Litmanen, P. Hokkanen and M. Kojo (eds), *Ydinjäte käsissämme. Suomen ydinjätehuolto ja suomalainen yhteiskunta* (Nuclear Waste in Our Hands: Finnish Nuclear Waste Management and Finnish Society). Jyväskylä: Sophi, pp. 43–64.
Satakunnan Kansa 13 December 1994, 14 December 1994, 30 January 1996, 29 March 1998, 5 June 1998, 27 May 1999, 19 August 2004. A regional newspaper.
Sisä-Suomen Lehti (1995) 16 November. A local newspaper.
Sunell, M. (2004) Suomalainen ydinvoimapoikkeus (The Finnish Nuclear Power Exception). In M. Kojo (ed.), *Ydinvoima, valta ja vastarinta* (Nuclear Power, Power, and Resistance). Helsinki: Like, pp. 179–207.
Sundqvist, G. (2002) *The Bedrock of Opinion: Science, Technology and Society in the Siting of High-Level Nuclear Waste.* Dordrecht: Kluwer Academic Publishers.
Suominen, P (1999) Ydinjätepolitiikan muotoutuminen Suomessa. (Formation of the Nuclear Waste Policy in Finland). In T. Litmanen, P. Hokkanen and M. Kojo (eds), *Ydinjäte käsissämme. Suomen ydinjätehuolto ja suomalainen yhteiskunta* (Nuclear Waste in Our Hands: Finnish Nuclear Waste Management and Finnish Society). Jyväskylä: SoPhi, pp. 15–42.
TVO (1980) Undertaking. The Letter from Teollisuuden Voima Ltd to the Eurajoki local government on 31 March 1980.
TVO (1995) *Olkiluodon Uutiset* (Olkiluoto News). October 1995. Information leaflet of Teollisuuden Voima.
Vahekoski, R. (1992) Eurajoen ydinvoimahanke jälkikäteen muisteltuna (Recalling the Nuclear Power Plant Project in Eurajoki). In *Our Home District Eurajoki IX.* The joint publication of the Eurajoki Folklore Society and the cultural board of the Eurajoki municipality, pp. 69–88.
Vieno, T., A. Hautojärvi, L. Koskinen and H. Nordman (1992) *Käytetyn ydinpolttoaineen loppusijoituksen turvallisuusanalyysi TVO-92* (Safety Analysis of Spent Nuclear Fuel Disposal). Technical Research Centre of Finland. Nuclear Engineering Laboratory. The Nuclear Waste Commission of Finnish Power Companies. Report YJT-92-33. Vantaa, Helsinki: Imatran Voima, Teollisuuden Voima.

7

The Temporary Nature of Societal Risk Evaluation: Understanding the Finnish Nuclear Decisions

Tapio Litmanen

At the beginning of the new millennium the Parliament of Finland made two historical decisions. First, in 2001 the Parliament accepted the plans of the nuclear waste company Posiva to proceed with their final disposal project for a spent nuclear fuel (SNF) facility in the municipality of Eurajoki. Following this, in 2002 Parliament made a favourable Decision-in-Principle (DiP) on the proposed project of the nuclear power company TVO to construct a fifth nuclear power plant in Eurajoki. Both of these decisions can be regarded as unique, because Finland is currently rowing against the current. The planning of nuclear waste management in other countries has come to a halt, or their important decisions have been postponed as the result of political disagreements and some countries are shutting down and decommissioning their nuclear reactors because nuclear power is no longer perceived to be an acceptable source of energy.

The intention of this chapter is to analyse the factors affecting Finnish nuclear decisions.[1] In the first part of the chapter the theory of societal risk evaluation is developed. Mary Douglas' cultural risk theory and its deficiencies are dealt in order to develop adequate explanatory tools to analyse the case. The second part of the chapter will deal with those cultural factors which characterize the Finnish mentality and which have to be included in any attempt to understand the way in which people perceive the risks of nuclear power. For instance, Finns have a belief in enlightenment thinking and are supporters of the welfare state's services and bureaucratic systems, which are seen as ensuring fairness in the making of important decisions. The overcoming of determinism and universalism in such structural cultural explanations is important, because those very cultural factors which affected the nuclear decisions of 2001 and 2002 had also existed in 1993, when the Finnish Parliament had rejected the plans of the nuclear power company to build a fifth nuclear reactor. Thus in the third part of the chapter a greater emphasis is placed on historical changes in Finnish society and the changes in the international circumstances in which the decisions were made. In order to understand the temporary nature of societal risk evaluation, we must consider not only internal factors such as the role of the national media, technology policy and lobbying, but also external factors such as business fluctuations, international agreements and global political opinion.

Societal evaluation of risk

To start the analysis of the temporary nature of societal risk evaluation we can take a look at the cultural risk theory developed by Mary Douglas. Her cultural risk theory has greatly furthered our attempts to understand risk as a sociological problem. As Lupton (1999a, 37) writes, Douglas is trenchant in her critique of cognitive and other techno-scientific approaches to the understanding of risk. Douglas argues that it would be naïve to think that every individual is a prisoner of his/her own cognitive scheme. We simply cannot overlook the social circumstances of human life. A person dealing with risks is not cold, calculating and selfish, and does not always pursue her/his own interests. That is to say, family, friends, and many other external forces have an effect on an individual's risk calculations. The erroneous assumption of some risk analyses has been the idea that the public response to risk is entirely individualistic. In some cases the public perception of risk is treated as if it were the aggregated response of millions of private individuals. According to Douglas (1992, 40), among other well-known fallacies of aggregated choice is that it fails to take account of individuals' interactions with one another, their advice to one another, their persuasions and the inter-subjective mobilization of beliefs.

The idea of the societal evaluation of risks, which is tested and elaborated in this chapter, is founded on the epistemological premise that individuals and institutions process risks and that most of the processing takes place in broader sociocultural, political and economic contexts. This evaluation and selection process is a normal part of the functioning of every single society, but there is always the possibility of cultural bias or distortion (Douglas, 1986, 91–101; Bauer, 1997).

Douglas (1986) provides several examples of cultural bias in organizations. One example is the theory of diffusion of responsibility, which involves the basic assumption that groups take greater risks than individuals because there will be more people to share the responsibility when something unexpected happens. Furthermore, all organizations have loyalty structures, which may be exposed, for instance, when a professional group or organization sorts out the causes and effects of a scandal. Professional organizations are usually concerned about their reputation; hence they seldom criticize their members; rather, they tend to protect them. Occupational loyalty also plays a large part in all organizations, i.e. there is a strong reciprocal loyalty inside all occupational groups. Moreover, the perception of moral issues by each individual employee directs the perception of risk within organizations. The more institutions are dependent on personal commitment, the more the risks are viewed from the perspective of fairness ('How fair are the decisions?'). For instance, risk was less tolerated in a workplace where the workers felt that they were being exploited. There are also many cultural and social variations in the fairness discourse. For instance, ordinary workers perceived fairness in terms of treating each employee equally, while line managers and managers saw fairness in terms of acknowledging the skills of each individual (Douglas, 1986, 66–7). Thus, risk taking and risk avoidance, shared trust and shared fears, are all dialogues on the arrangement of social relations. Mary Douglas (1986) poses the question: How do we select the risks that we take seriously? Her answer is that we choose risks at the very same time as we choose social institutions – that is, institutions and the values of individuals are important factors in risk perception. The logic is that an individual cannot look in all directions simultaneously; therefore, people arrange their lives around a social perspective.

At the societal level, we find that every form of society produces its own, selective view of the environment, and this is reflected in the kinds of issues that are considered to be dangers. Every society develops a unique system of responsibility, which observes certain hazards. Some dangers are collectively underestimated or forgotten, whilst others are considered to be worthy of attention and are brought into the public debate. According to Douglas and Wildavsky, societal criticism is voiced mainly in situations where the trust in the societal control of dangers has disappeared and fear has become the dominant factor. In such situations, criticism may serve as a way of identifying and changing issues that are believed to promote both fear and the disappearance of trust. In other words, the principal targets of societal criticism denote the most important social anxieties (Douglas and Wildavsky, 1983, 7).

Consequently, it may be argued that societies, social institutions, social groups and individuals are continuously classifying and arranging information about dangers. Hence, we are continuously assessing, singling out, and identifying certain phenomena as risks, a process of which we must be aware. For instance, Lupton (1999a, 13–14) has identified six major categories of risk that currently predominate in the concerns of individuals and institutions in western societies: (1) environmental risks; (2) lifestyle risks (related to consumption); (3) medical risks (related to medical care or treatment); (4) interpersonal risks (related to intimate relationships); (5) economic risks; and (6) criminal risks. The introduction of these risks to the political agenda by different actors usually also involves the requirement that society should try to eliminate the causes of these risks, take measures to minimize these risks, manage these risks better, or develop ways in which to ensure more equal distribution of these risks.

In many respects Mary Douglas' cultural risk theory bears a close resemblance to typical post-structural anthropological theories (see, for example, Milton, 1996, 89), or typical functional structuralist analyses of sociocultural phenomena (Lupton, 1999a, 56). However, Douglas' ideas differ somewhat from the ideas of most other post-structuralists, because her idea of the relationship between culture and social organization is essentially deterministic rather than dialectic in nature. Many post-structuralists emphasize individual freedoms, such as the freedom of choice. Douglas emphasizes the deterministic power of social organizational forms over the individual, who is reduced to a passive perceiver of risk. She claims that each form of social organization produces a specific type of perception of the world.

According to Milton (1996, 89), formulations such as that presented above illustrate the influence of Durkheim on Douglas's theory. Different social organizations produce different rules for interpreting the world, different cosmologies, and different cultural approaches. Highly homogeneous societies and communities produce uniform approaches and ways of understanding, whereas more complex societies and containing a greater number of different elements produce more heterogeneous approaches.

According to Milton (1996), Douglasian cultural theory has a number of problems. Milton finds particular problems in the way in which the forms of social organization produce or determine a cultural approach. In Douglasian cultural theory, the underlying idea is that a person's cultural perspective changes constantly as a consequence of changes in social organization (Douglas and Wildavsky, 1983, 192). Furthermore, Douglas and Wildavsky claim that social change is primary, while cultural change is consequential (Milton, 1996, 98–103).

Even though Douglas's theory helps to overcome some problems in the analysis of societal risk evaluation, it is based on a relatively static picture of culture, and the idea that the freedom of an individual is rather limited. The static structuralist emphasis is, for example, visible in the way the culture behind the risk perception is determined by social organization.[2]

According to Douglas, culture is a commonly shared collection of principles and values which legitimize a person's behaviour. Douglas argues that the principles and values of people are not developed in isolation; rather, many of them are directly influenced by public institutions. This emphasis entails that individuals are considered to perceive risks in a relatively passive way, whereas some scholars strive to cover such questions as citizens' active participation in the social construction[3] of risk and the definitional struggles over risk (e.g. Litmanen 1996; Lidskog and Litmanen 1997). We can replace the deterministic overtones[4] of Douglas's cultural theory with a more voluntarist emphasis. But what is missing in this type of analysis is any consideration of the explicit role of powerful societal actors and institutions, such as government, media, science and (lobbying) pressure groups, which can certainly affect people's perception of risks and whose activities may produce variance in the evaluation of societal risk. Not only can they reflect the current risk perception but they can also produce drastic changes in society, which may even have a causal power over either individual people's risk perceptions or the predominant societal risk assessment. Thus, the research orientation of this chapter is cultural risk theory, which has also been elaborated to cover contextual factors and societal processes. In order to understand the complex dynamics of a societal risk evaluation we have to consider the contextual and embedded nature of individual and collective interpretations, which means attaching them to the relevant societal, historical and geographical contexts.

Faith in enlightenment, state, technology and bureaucracy

If we take seriously Douglas's idea that societies create their unique risk folders and that social institutions play an important role in affecting how people assess the seriousness of different kinds of risks, we must turn to an analysis of what is culturally characteristic in Finland and which are their most important societal values.

Depending on the authors, the picture of contemporary Finland varies, but one common feature appears to be a quite unique feeling of trust among Finns. There are several ways to characterize contemporary Finland, but in this case it may be enough to refer to Pesonen's and Riihinen's (2002) analysis of the Finnish political system, where they stress that one of the basic preconditions for Finland's economic and social development has been political stability, which itself is based on a high level of social capital. In summary, people trust each other, corruption is rare and social morals are high.

Although this type of social capital has decreased during the first years of the new millennium, it is clear that the level of individual trust is still very strong (Ilmonen and Jokinen 2002, 195–215; Eurobarometer 62, 2004). Generally speaking, trust is a socially binding, integrating mechanism, which both creates and sustains solidarity among the members of a community and increases and diffuses the collective feeling of security. A national atmosphere of trust contains such contributory factors as faith in enlightenment (Pesonen and Riihinen 2002; EVA 2001), faith in state (EVA 1999a, Hautamäki 2001, 37; Castells and Himanen 2002, 135), faith in technology (EVA 1999b, 2001; Castells and Himanen 2002, 133) and faith in bureaucracy (Hautamäki 2001, 37). These important cultural factors affecting Finnish nuclear decisions are inextricably linked.

Faith in enlightenment has enabled the development of the nation-state from the nineteenth century onwards. There have been no discontinuities in the Finnish faith in enlightenment values (Konttinen 1991, 180–216, 252). Present-day surveys indicate that the population agrees with the authorities that one success factor is the Finnish investment in education, research and science. For instance, the results of the OECD's comparative studies have shown that Finland is among those top countries which invest most in their education system. A strong comprehensive school system and the strength of a free education system are supported by the population, as nearly 80 per cent of people share the view that the Finnish success story of the 1990s was due to a high level of education and equality of education opportunities (EVA 2001). Science and research have also been regarded highly both among laymen and authorities (Tiedebarometri 2001, 2004). Legitimation of the significant financial support for science, research, universities and other higher education institutions can be found from the consensual spirit among people. Science and education are perceived not only as success factors in international competition but also as important cultural values, which characterize and construct national identity.

Interestingly enough, the faith in enlightenment is connected to faith in the state (Konttinen 1991, 79, 265–6). For instance, attitude surveys conducted in the 1990s illustrate convincingly that the majority of Finns are advocates of a welfare state and comprehensive welfare services. The economic boom at the end of the 1980s, which was followed by a period of rapid decline at the beginning of the 1990s, is one explanation for the strong levels of public support for a welfare state. Despite a fairly recent period of general prosperity in the 1980s, which led to a temporary decrease in the level of faith in strong state structures, an increase in the criticism of social services and a series of protests against state paternalism, the trust in governmental institutions is also very high in international terms. One characteristic of Finland and other Scandinavian countries is that they have a unique trust in the state, institutions and authorities. This can be regarded as one factor which enables difficult societal decisions to be taken without people feeling either excluded or that they have been deceived.

One has to admit frankly that the trust in politicians has been on a downward trend, but we have to make a distinction between party politics and institutional structures, which support the thorough preparation governmental bills and implementing governmental decisions (cf. Ilmonen and Jokinen 2002, 244). According to Eurobarometer 62 (2004), the most trusted institutions in Finland after the army (with 93 per cent of Finns rating it as trustworthy) and the police (91 per cent) was the Finnish legal system (73 per cent), the Parliament of Finland (68 per cent) and the Government of Finland (67 per cent). Comparison to politicians indicated the importance of this distinction, as the same Eurobarometer revealed that only 26 per cent of Finns trusted the political parties and as many as 70 per cent say they tend not to trust them.

Finland has been regarded as a prototype of the information or information technology society, but there has been less interest in those scholars, which explains the economic success of the country from the point of view of the functioning of the labour markets. The so-called January Engagement in 1940 began the favourable development of Finnish corporatism, under which the trinity of government, the labour unions and employers' organization produced peaceful conditions in the labour markets, brought predictability to societal development and was able to foster economic development.

A strong state, corporatism and bureaucracy are quite commonly perceived as suppressing the individual, but recent feminist studies have indicated that they can also be regarded as friends of women or individuals in general if they operate effectively, are open to citizens' opinions and listen to the current needs of the layman. Historically, the social distance between the citizens and the authorities has been quite close in Scandinavian welfare states (cf. Konttinen 1991, 181–2). For example, some factions of the anarchist movement value the welfare state very highly as they receive financial aid from the state or from the municipality and some movement activists enjoy liberty because of the unemployment allowance they are paid.

It is more difficult to account for the faith in technology, which over the course of the past decade has become almost a mania for new technology or fanaticism about mobile phones. On the one hand, the basis of faith in technology is the above-mentioned enlightenment tradition, which has become embodied in strong support for science and research and development work. On the other hand, the underlying force is the history of the industrialization of Finnish society. The ability to rapidly adopt new technological innovations has been characteristic, for instance, of the wood processing industry, which developed in rural areas in the nineteenthcentury because of the access to water (e.g. Kerkelä 1996). Small communities grew into bigger cities as the new paper mills created jobs for the non-landowning population of the region. This meant that prosperity and rapid societal development were associated with technology and industry. The recent success story of the electronics and information technology industry has strengthened the faith in the omnipotence of new technology as the mobile phone company Nokia and other similar enterprises lifted the nation out of economic crisis and put it on the path to economic growth. Thus, Finnish industry received a vote of confidence from the people as it has managed to have a positive influence on societal development at the turn of the millennium.

In addition to the Finnish faith in technology one has also to remember the faith in administration because people support the structures, mechanisms and procedures of the welfare state. The maintenance of the welfare state system is accomplished by sometimes almost invisible administrative work. In the Scandinavian countries the public sector is well developed and the administrative system is thought to operate for the good of the citizens as a large number of governmental organizations and officials follow the rules and ensure that the procedures are correct. Although the further elaboration of Max Weber's theory of bureaucracy suggests that the superiority of rational bureaucracy knowledge and the monopoly-like role of legal discourse is challenged not only by capitalist entrepreneurs or market forces but also by intellectuals and well-educated laypeople who are able to oppose power configurations with knowledge available in societies (Stehr 2001, 4–5), the Nordic welfare states' bureaucratic administration resembles the Weberian model of an ideal bureaucracy, which derives from impersonal legal norms, where the authority of the administrative apparatus is based on technical knowledge and where the continuous procedurally correct work is carried out by officials with clearly circumscribed spheres of competence (Stehr 2001, 60–2).

An illustration of this kind of trust in bureaucracy can perhaps be seen in the national nuclear waste management system. The international nuclear industry and the international organization working in the field of nuclear power have claimed that it is worthy of great praise, because of the management of the procedure and the excellence of the administrative work. The main principles of the technical solution itself were copied from Sweden, but the implementation of the twenty-year long procedure to the final Decision-in-Principle of Parliament in 2001 to dump high-level nuclear waste in Eurajoki has been praised as a magnificent administrative operation. It was recognized that an enormous amount of effort was put into the management of the process in changing social and political circumstances.

Flexibility and openness to new initiatives was combined with strict timetable structures so that from 1983 onwards the information on the dates of the most important administrative and governmental decisions was known to all of the different parties and stakeholders (Vuori 1997, 177; Rasilainen 2002, 174). The opponents of nuclear waste management described their frustration with the procedure by stating that the process is like a train making headway on the railroad, stopping only at certain points but without junctions, passengers travelling on single tickets only and engine drivers ensuring that it arrives on time at its destination (Rosenberg 1999, 277–80; 2004, 169–70).

The fundamental question to be considered here was what made it possible to achieve two positive nuclear power decisions in Finland in 2001 and 2002. Tentative analysis has provided evidence that cultural factors played an important role in trying to understand these decisions. However, cultural factors cannot provide a sufficient explanation of this process. The political history of nuclear power in Finland does not support the hypothesis that only cultural factors have affected the societal acceptance of nuclear technology. The previously described cultural preconditions existed in 1993, but the result of parliamentary decision making was negative. Only a decade later the Finnish Parliament was looking at nuclear power in a very different light. Nuclear power was perceived as a suitable source of energy and the Finnish Parliament gave permission to proceed with the planning and construction of the fifth nuclear reactor.

Expressed in more sociological terms, the structurally oriented cultural explanation does not yield a satisfactory description for these three cases of nuclear decision making. Therefore one has to elaborate such theoretical tools and mitigate the strong cultural structuralism by emphasizing more contextual and procedural factors. What is required is a more sophisticated analysis of local particularities and descriptions of certain general societal processes. This chapter deals with a number of contextual-procedural factors: 1. The state of political networks, 2. The altered media circumstances, 3. The scientization of lobbying, 4. The profound change in the relationship between science and society, 5. The neutralization of environmental arguments, 6. The changes in values and societal structures, 7. Reforms in national competition strategy. The assumption is that over the period 1991–93, the period of the previous nuclear power debate, these factors were all very different from those that prevailed during the most recent nuclear debate in the years 2000–02. The analysis of these national peculiarities is followed by a conclusion in which the explanatory scheme is extended to include recent international developments, to overcome the evident parochialism of the chapter and to point the future direction of the study of factors affecting societal risk evaluation.

The persistent construction of a pro-nuclear political network

In 1993 one distinctive characteristic of the political network was the differences of opinions among the political elite in relation to questions elating to nuclear power. Within the political parties there was a considerable degree of opposition to nuclear power in general and to the idea of the construction of a fifth nuclear reactor in particular. For instance, the right-wing government was divided over this question and Prime Minister Esko Aho (The Centre Party, Suomen Keskusta) brought the decision into parliamentary proceedings, although he was personally opposed to the idea. This was characteristic for the proceedings and enabled the opponents of nuclear power to reject the initiative.

At that time an anti-nuclear or sceptical attitude was widespread among many of Finland's political parties. They had been receptive to the messages of the anti-nuclear movement of the 1980s, because of its association with the electorally popular stance of environmentalism, and furthermore the Chernobyl accident was still fresh in the public's memory. In addition, the collapse of the Soviet Union highlighted the issue of nuclear waste management and affected the public debate because the other nuclear power company, Imatran Voima Ltd (IVO), exported their high-level nuclear waste – first to the Soviet Union and, later, to Russia. Thus, at the beginning of the 1990s the new nuclear power project ran into difficulties.

When we compare the Finnish political climate of the 1990s to that nearly ten years later, it can be seen clearly that by this time the Green League had become the only government party to maintain an anti-nuclear stance. Both the Social Democratic Party and the National Coalition Party were unanimous in promoting the construction of a fifth nuclear reactor. The consensus between these two major parties in government did not offer the opponents of the project such good opportunities to generate political debate as had existed in 1992 and 1993 (cf. Nissinen 2004). It can be suggested that on this occasion the opposition to nuclear power had become stigmatized even before the parliamentary proceedings, despite the activists' attempts to avoid this occurring. The activists insisted that the anti-nuclear movement was not only the action of Greens, radicals and fanatics, but they did not succeed in convincing the other stakeholders or the public (Hylkilä 2003, 101–3). Before the final vote in Parliament the parties did not want to make a group decision and the matter was left entirely to their representatives' discretion. On the construction of more nuclear power, the Left Alliance voted 16 against and 4 in favour, the Swedish People's Party 10 against and 4 in favour, the Centre Party 22 against and 25 in favour, the Social Democrats 20 against and 32 in favour and the Christian Democrats 6 against and 4 in favour.

The construction of a pro-nuclear political network had already begun in 1995, only two years after the negative decision. Hylkilä (2003) has made a study of the lobbying for nuclear power, conducting interviews with many of the key actors. He found that there were two distinct phases in the construction of the pro-nuclear network. For the first five years, from 1995 to 1999, the lobbying was carried out by the nuclear power company TVO and some industrial and commercial organizations. They held a series of information and public relations events to which members of Parliament were invited. In the second phase, following the parliamentary elections of 1999, lobbying became more systematic, after the group of members of Parliament developed a strategy that lobbying should only be undertaken inside Parliament. As a result, the four MPs established an internal lobby, which even coordinated the work of the lobbyists of interest groups. Without the permission of this inner core other lobbyists were not allowed to do anything. Each of these MPs was from a different party, so that they were able to promote their aims within the Social Democrats, the Centre Party, the Coalition Party and the Left Alliance. By the Spring of 2002 the internal lobby group inside Parliament consisted of 20 MPs from different parties. In constructing the pro-nuclear network the group concentrated their powers of persuasion on those MPs who were wavering in their views or who were opposed to nuclear power. They had a twofold strategy: on the one hand converting individuals by analysing the political backgrounds of individual MPs and then developing suitable tools for each one; and on the other, designing more general solutions to national energy policy. One efficient method was to address the constituents of each MP and to use them as auxilliary lobbyists. In convincing the MPs collectively the lobby group tried to make a favourable decision more attractive by planning a supportive programme, which would also lead to an increase in investment in renewable energy sources. It is noteworthy how the lobby group arranged nationwide official mettings with many stakeholders (such as local energy enterprises) around the country. The most crucial question at these meetings was to transform the pro-nuclear ideology into more factual news and to indoctrinate the audience with their views on energy policy.

The nuclear industry feeding the journalist with information

Changes in the national media and mass communications also helped to pave the way for a positive nuclear power decision. In comparison with the previous nuclear debate in 1993, the messages from the environmentalist movement were considered to be relevant and the journalists were much more favourable to the bulletins and campaigns of the movement. Tommola (2001) has pointed out that in reporting the nuclear waste issue the journalists were reconciled to being fed knowledge by the nuclear industry and they had become increasingly fed up with so-called alternative perspectives.[5] The newsmen themselves reported that they could not find reliable sources of information on nuclear waste issues among the opponents. The MPs interviewed by Hylkilä (2003) emphasized that, in general, the media was much more sympathetic to the nuclear industry than had been the case during the previous debate in 1993. This view was both corroborated by supporting and opposing MPs (Hylkilä 2003, 100–5, 147–50).

It is appropriate to conclude that there had been changes in the way in which journalists perceived nuclear power issues. In seeking an explanation for this change, one cannot rely on the first and most obvious explanations – that they were acting collectively and that their ability to perceive the issue was deficient. We have to consider why they were acting in a collective manner. Again, this shows the limits of a culturally oriented explanation, as will be shown in the analysis of the institutional circumstances of journalists.

Today the structure of the mass media is very different. In addition to newspapers, there are also other channels of communication and during this period the characteristics of conventional electric distribution channels of information have been in a state of constant change. The number of radio stations and television channels has multiplied and the news is regarded highly. Stations and channels are improving their image by investing in basic news services and news broadcasts, which do not necessarily need deeper analysis or investigative journalism. On the one hand, the proliferation of news has become entangled with real-time reporting and on the other hand the contents have been brought closer and closer to entertainment (Ruostetsaari 2003, 124).

The changes in the structure of modern media have meant that powerful media trusts have become more powerful as they have adopted new channels and means to reach the public, but at the same time this trend has not led to an increasing pluralism in communication. On the contrary, it has decreased the level of pluralism in the mainstream media. There has been the formation of media syndicates, together with the creation of new information channels. The multi-channel communication strategy of media companies is interwoven with the idea that diversity among the audience can be reached with new forums, which reach the target audience. For instance, traditional mass communications, like newspapers have closed their ranks and established Internet-based newspapers to have many more channels to serve the audience (Nieminen et al. 1999, 1–7). It sounds paradoxical, but it may be true that the increased variety and competition within the media have caused narrow-mindedness in the contents of media, i.e. more media and communication has not increased the level of pluralism (Ruostetsaari 2003, 124).

At earlier stages nuclear power symbolized the era of the environmental movement and was the generator of upsurges of feelings, but today the importance of nuclear power is viewed differently. The increasing volume of multi-channel news transmission by media companies is devaluing the importance of single topics, such as those surrounding nuclear power. Furthermore, ways of life have brought changes to the audience of mass communication. The short attention spans of both listeners and television viewers have encouraged the development of a more rapid news transmission. The age of undifferentiated audiences has been left behind as the television channels specialize and newspapers develop their profiles to correspond to their desired target audience. As a result, news items now have to be particularly dramatic and significant. It may well be the case that nuclear power is no longer regarded as special.

One MP who was also an active opponent of nuclear power claimed that a sympathetic disposition towards nuclear power could be seen from the way in which journalists reported on nuclear power in terms of an important societal decision made by sensible and mature adults (Hylkilä 2003, 148). Another MP, an active supporter of nuclear power, shared the view that the media's general perception of nuclear power had changed and said in the research interview that it was not fashionable to write critical reports about nuclear power (Hylkilä 2003, 148).

There have also been changes in the journalistic profession. In recent years a younger and better-educated generation of journalists is gradually coming to the fore. With their university education they are better able to understand, process and in some cases even to produce scientific studies. A significant part of their everyday work is to write scientific reports for the news and to popularize scientific research results for public consumption. They are workers of the information society, where the air is thick with rather different and heterogeneous information, where the cultural influences and values vary from group to group and where there are continuous struggles over meanings between the parties.

One can surmise that as the older generation of journalists was more eager to take a moral stance to different issues and interpreting the profound sentiments of the public at large, the younger generation of journalists emphasizes expert knowledge and uses the evidence from scientific papers when constructing their reports. Scholars of journalism have claimed that one tendency in contemporary journalism is neutrality, and seeing matters from a distance. The mainstreaming and factualization of issues are typical strategies with the effect of achieving a neutral approach to matters. Raittila (2002, 76) made a study of the reporting of nuclear waste management in the media. He found that the nuclear waste company Posiva did not have to advertise their management model so much, because in the reports it was neutralized by being presented through timetables and technical solutions. Journalists also compared the model to natural processes and other innocuous projects. The factualization of the news was manifest in how journalists did not mention the representatives of the company as speakers or sources of information although the material would have allowed this (Raittila 2002, 77). In his study, Raittila described how journalists did not reveal their sources of information, but presented the facts and arguments from the material produced by the company. This caused the phenomenon of factualization, i.e. the arguments of one party became unbiased facts as the journalist processed them into news.

In the past journalists were often forced to adopt a stance on matters because there was less information available; today, this is more difficult because the communications professionals and the lobbying organizations supply suitable messages and packages of information. For instance, Raittila (2002, 77) has written that '... Posiva's management of publicity was active and professional. The aim of this kind of management is not very often to be seen in the media, but ensuring that the company's own view is relayed to the public. In their own bulletin Posiva let the local residents speak for the final disposal project.' The contrast between Posiva and the environmental movement or local opposition groups is a particularly strong one. The relatively poor level of the public relations and information work produced by anti-nuclear activists was simultaneous with Posiva's professional management of publicity, but the journalists were not receptive to alternative views. Both Raittila (2002) and Tommola (2001, 91–5) discuss the question, because the journalists interviewed by Tommola regarded it as a problem that in their nuclear waste management stories they could find no competent experts or reliable citizen activists to balance the information supplied by Posiva. On the other hand, Hylkilä (2003, 104) depicts the frustration of anti-nuclear activists when they realized that even authoritative independent individuals had difficulty getting their voice heard in the media and how journalists' stereotypical ideas about those opposed to nuclear power affected their choices of interviewees.

One may conjecture that the moral codes of the older generation of journalists were closer to the general moral tone of society. Changes in society and its institutions can also be observed in the media, where the professional standards of journalists may come not so much from other parts of society, but from the internal debate about the moral rights and obligations of the profession. The perpetuation and reproduction of the internal moral codes of journalists is also fostered by scientific research on journalism, institutionalized media critique and various professional education and training systems.

In addition to moral codes, which may today encourage journalists to distance themselves from societal struggles, to maintain their independence and to develop a reporting style resembling objectivity, the reporters have their own social networks. It would need a larger research programme to shed light on the networks of journalists, but in this case we can only refer to the work of a few scholars. For instance, Ruostetsaari (2003, 124–5), who conducted an investigation of Finnish elite and power structures, argues that the commercialization of the mass media has brought economic interests closer to the everyday work of the media and that journalists and economic actors have much better relationships than had hitherto been the case. In addition to this, Perko (according to Ruostetsaari 2003, 125–7), himself both an editor-in-chief and a scientist, detected that some journalists derive perhaps an excessive pleasure from being at the core of political, economic and cultural power. The identification of journalists with power structures is a result of harmonious coexistence and co-operation (Ruostetsaari 2003, 229–31). This phenomenon reflects the overall change in society, which has trained citizens to be capable of managing the complex processes in varying organizational circumstances – which used to be called workplaces. Thus, the construction of the social networks of journalists and their assessment of expert knowledge is much more of a structural than a cultural phenomenon, and cannot be reduced to the simple actions of individual actors. The symbiotic change of mass media goes hand in hand with changes in the other parts of society.

The scientization of lobbying

When comparing the debates on nuclear power in the 1990s and at the beginning of the twenty-first century, it is evident that the degree to which science has penetrated the public debate has grown significantly. The scientization of the debate prior to political decision making differs from that of former days, when the basis of any appeal to the public was the worldview or ideology which contributed to the discussion about the values behind the decisions. Today people do not regard ideology so highly, because in appealing to the public one has to take advantage of science. The parties have to roll out deliberate scientific knowledge to convince the audience and to be newsworthy, but also to be competitive in gaining the attention of journalists. In the promotion of their aims the organizations have to advance a strong scientific case. For instance, it is no surprise that major environmental organizations have their own scientists or resources to fund the necessary research.

Communications scholars discuss the mediatization of society, meaning that societal institutions are becoming dependent on the ways media function and on the image of their work conveyed by the media. This assumption is based on the idea that other societal institutions are trying to increase their importance in adapting to circumstances where the centrality of media is crucial and where their own functions have to be adjusted to correspond to the functional logic of media (Nieminen et al. 1999, 1–7; Väliverronen 2002; Schulz 2004). But then again, we have to bear in mind that the sociology of knowledge has investigated how the scientization of politics and the politicization of science is an undeniable fact. We can conclude that mediatization is only a partial description of a development in which the subsystems of society are developing to correspond to the altered circumstances. The argument about the mediatization is an overestimation, because it focuses on the way in which the subsystems of society develop their functions to correspond to the logic of media and tend to forget the importance of other institutions, like science and politics. In the light of Finnish nuclear decisions it appears that what is currently under way is a more fundamental intensification of symbiotic relationships between politics, science and media.

Before both nuclear decisions, in 1993 and 2002, the level of parliamentary lobbying was intense. Both the pro- and anti-nuclear parties had learned the lessons of 1993 lobbying, when the pro-nuclear party tried to convince the MPs through aggressive lobbying techniques (Kyllönen 2004; see also Ruostetsaari 1998, 158–9; Hylkilä 2003, 45). After their defeat the MPs and also the pro-nuclear parties concluded that the lobbying campaign was a fiasco. Following these experiences both parties realized that the political culture and the national modus operandi differ from those political cultures, permitting aggressive sway by authorities and opinion formation by politicians. A much more suitable way to promote the views is restrained and respectful opinion formation seasoned with scientific facts, tactful behaviour and trust in the normal functioning of institutions.

Both the management and labour organizations carried out efficient pro-nuclear information campaigns, giving the impression of the scientization of lobbying. The parties produced convincing research reports and information leaflets, which led to the drawing of clear conclusions. They stressed that Finland needed nuclear power for a number of reasons: (1) because the consumption of energy was increasing; (2) because otherwise Finland would become even more dependent on energy from Russia; (3) because Finland was committed to reduce greenhouse gases; (4) because without the fifth nuclear reactor the competitiveness of Finnish industry would collapse; (5) because of a positive nuclear decision Finland would be able to invest more in the research and development of renewable energy sources; and (6) because this was the only way Finland could secure the structures of the welfare state. By contrast, the anti-nuclear camp presented scientific expertise and views which had a basis in science.

The attention of the decision makers was gained through discreet and convincing scientific-looking information. The legitimation of the decision was not reached by appealing to a commonly shared moral understanding or personal political views; the decision had to be supported by considered fact. In addition to this, the decision makers were forced to discuss the reliability of sources of knowledge, in more detail, likewise the experts to give their views in the preparatory stage (Lammi 2004).

Intensification of the social responsibility of sciences

Over a very short period of time Finland's industrial structure had changed from being focused on raw materials, energy and capital-intensive industries towards production characterized by knowledge and know-how. Over the same period, it had reached top rankings in comparative tables of international competitiveness (Maula et al. 2004, 281). Its international standards of science were also rated very highly. For instance, measured in terms of the number of publications and citations in international journals, the Finnish science community is ranked as one of the ten most productive of the OECD countries; in terms of the proportion of scientific publications to the size of the population or the gross national product, the ranking was even higher.

Increasing efficiency, international competition and the intensification of the social responsibility of the universities has dragged universities into societal processes and particularism.[6] One obvious trend is towards the development of close relationships with industry and commerce and towards a responsiveness to the various expectations and needs of society. A means to suppress 'unfruitful' academic criticism is to advance a claim that the universities must reorganize their research activities in order to empower science to promote constructive advances in the general interest.

Ruostetsaari (2003, 128) notes that publicly funded research and development activities are moving closer to the areas of administration and business. The increased dependence on finance from outside the universities or the state budget has created new kinds of loyalty relationships. The commercialization of science is a potential research topic of its own, but we can refer to the work of Ruostetsaari (2003) when he states that during the 1970s and 1980s the universities moved from a planned economy to a commercial orientation. The universities have become increasingly committed to productive interests and commercial considerations have become more and more important to them. Today even scientists have to think about advertising, publicity and a nice clean image.

Stehr (1994), Gibbons et al. (1994) and Nowotny et al. (2002) analysed the relationships between science and society. Independent of the authors, the common notion is the disintegration of borders between science and society. For instance, Nowotny et al. (2002) emphasize that scientific communities have become socially more open and all-embracing, which has in turn made science more socially robust. The intensification of social relationships between science and other societal actors creates situations in which interests, values and even norms are submitted to negotiations, struggles and decision making. The assumption of an intensification of science's relationships casts doubt on conventional epistemological ideas. Because of the new role that science is playing in society, should we only accept the relativity of knowledge as a starting point for scientific research? Or how should we understand the philosophical demarcation problem when the distinction between scientific and non-scientific knowledge is fading away because of the dispersion of scientific institutions, the proliferation of the social usage of scientific facts and the merging with of research results into societal processes? Bertilsson (2003) describes how science and entertainment are united into infotainment in the media. For instance, with the help of information technology the visualization of new scientific research results is produced and transmitted to audiences in the form of entertaining documentaries.

The intensification of the relationships between science and other societal actors could also be seen in the societal processing of the nuclear power decision. The opponents of nuclear power were not able to find 'the right kind of scientific expertise' among natural scientists. It would appear to be the case that the anti-nuclear movement can no longer get support from scientists with critical attitudes to nuclear technology. In point of fact, Tammilehto (1996) documented how during the 1970s and 1980s the anti-nuclear movement had a pool of experts from natural and engineering sciences to turn to as a source of public comments or statements. Presumably the stepping up of competition, claims for making results and the abovementioned intensification of societal relationships have decreased the enthusiasm of scientists to make political or moral statements. Both natural scientists and also social scientists have become increasingly cautious about making political statements or critical analyses of societal processes.

The dissolution of old front lines

The deep structure of the recent nuclear debate was characterized by two factors. Firstly, the magnitude of the exploitation of science and research in argumentation was greater than during the preceding nuclear debate. Every party took advantage of the high status of science on a scale which meant the strategic use of science to further political aims. Secondly, it was much more difficult to distinguish between the parties because they had learned from the experiences of the previous nuclear debate. In 1992–93 the front lines between the pro- and anti-nuclear parties were more clear-cut: the arguments of the pro-nuclear camp were rooted in 'hard' economic values and prototypical industrialization models, whereas the argumentation of the anti-nuclear movement was based on so-called soft values, with priority being given to nature conservation and life-saving values. In brief, in the course of the previous nuclear debate the division had been between industrialism and environmentalism, where the views of the parties were easy to anticipate, and the comments on the issue were frank and even straightforward. The most common pro-nuclear argument was: If the construction of a fifth nuclear reactor is allowed, economy and employment will be improved. The anti-nuclear camp perceived the issue from the point of view of short-sighted economic interests, which meant excessively high technological risks at the cost of nature and humankind.

In 2002 the style of argumentation was completely different. Hylkilä (2003, 153–4) stated that those MPs who had lobbied for nuclear power chose a concession strategy. This meant that they took into consideration the interests of their opponents, but also the present values of Finns. They softened the old lines of 1993, especially when they realized that in the conversion of those who were still unsure of their view more emphasis had to be put on the Kyoto climate agreement and the environmental obligations it entailed (Hylkilä 2003, 87). The nuclear industry and the actors close to it, such as the labour unions and the employers' organizations, started to produce scientific facts about environmental matters in accordance with their aims. In the pro-nuclear argumentation the national climate strategy and country's international commitments to reduce greenhouse gases were matters of great weight. The argument goes something like this: the further construction of nuclear power will enable Finland to reduce the emission of greenhouse gases and thus promote nature conservation. The peculiarity in the debate was that the anti-nuclear camp spent a great deal of time and resources on convincing the other parties that the construction of more nuclear power plants would not improve employment in practice compared to investment in the research and development of renewable energy sources and how the electricity produced in nuclear power stations was not competitive enough for the common Nordic electricity markets. This side also included scientific facts to refute the views of the other side and to disprove the arguments of the other party. The conclusion of the debate might be that the pro-nuclear movement was more successful than the anti-nuclear lobby in tempering their public image and their statements. The pro-nuclear parties claimed that nuclear power was an important factor in securing the continuity of the welfare state, but they also promoted the ideas that societies have to rely on nuclear power in advancing women's rights and addressing many of the problems faced by modern families. At the same time the opponents of nuclear power concentrated on employment and economic calculations and demanded increased investment in the development of alternative energy sources.

Impoverished Finland did not dare to take a nuclear risk

Despite the serious economic problems at the beginning of the 1990s the country did not dare to take any nuclear risks. Half a million jobs were lost, the country's gross national product was falling dramatically and there were apparently no limits to the increases in the national debt. Nevertheless, the politicians did not allow the private sector to make significant investments in nuclear power. The peculiarity of the historical developments is that following an economic boom and a time of prosperity the country was now more willing to increase nuclear energy production. The question is: how can we explain the change in views?

During the 1990s people's thoughts were dominated by fears originating in the depression and the turbulence of the changing world order. Opinion polls and attitude surveys indicated that people were fearful of what the future might hold. Thus the choices were dominated by a desire for security and for the minimization of risks. The nation went into its shell, re-invented family values, decorated their homes, paid attention to the education of their children, bought many more pets than ever before and were eager to speak more openly in public about private matters, such as bringing up children, relationships between husband and wife, love affairs and sex. Scholars of communication described how these developments were also seen in the media. Today this trend where intimacy goes public is rather commonly seen in web blogs, which sometimes are online diaries, where people publicize a running account of their personal lives. After taking a short breather with their personal affairs people were ready for a rapid increase of global social and economic interaction. At the societal level this can be seen, for instance, in how Finland joined the European Union on 1 January 1995 and how the internationalization of the economy occurred on a large scale.

According to Douglas (1986, 1992), we may theorize that at the beginning of the 1990s Finns were not eager to increase the already heavy burden of risk. According to cultural risk theory rich, affluent and educated people can afford to take risks, but poorly paid, badly off and uneducated people have no desire at all to increase their burden of risk. We can also refine this way of thinking to be more collective by stressing that the community of a nation which has already many risks and poor chances of protecting themselves rejects the change of taking more risks.

Such a generalization is anything but simple, because we also have cases in point which illustrate that the unempowered nations have to risk the most of all. One solution to this deficiency of cultural theory of risk might be to place an increased emphasis on contextual factors. For instance, a much better way of explaining[7] the negative nuclear power decision in 1993 is to point to the contemporary contextual factors. The rapid economic change from prosperity to economic austerity made both people and politicians cautious. In addition, both national and international economic and security structures were on the move.

Changing to a higher gear

When the economic structures both inside and outside the country began to change at the beginning of the 1990s, the key mission of politicians was to modify the welfare state to correspond to ever-changing international circumstances. Long gone are the days of a humble nation, as nowadays it is permitted to boast that Finland can be regarded as one of the most competitive countries with open markets and an insignificant level of regulation. In earlier days it had often been stated that 'Finland lives by its forests'. Today this slogan is no longer appropriate because of the well-developed information and communication technology sector and the birth of the information society. One essential precondition for the birth of an information society has been persistent and goal-directed development work, to which science, education and industry have all made contributions. The aim of consensual technology policy has been the promotion of knowledge and science-based economic growth. Over the course of the past two decades the priority of public research and development investments has been the development of the information technology sector. In spite of the recent uncertain business conditions in information and communication technology, the technology policy has been successful. This is shown by the unparallelled growth of the high-tech cluster and the rapid structural changes in the economy.[8]

With an improved level of international competitiveness and high rankings in various international comparative studies the nation's self-confidence has increased. Therefore the renewed national strategy is called a courageous technology policy,[9] meaning that the country dares to let go of the past and make bold plans.

Increasingly, societies are orientating themselves towards the control of their future through the development of scenarios and future studies. The aim is both to foresee coming events and to take control over the future in order to secure desirable advances. Among the indications of the coming of a future-oriented society are the increasing popularity of risk research (cf. Litmanen 2001, 36) and changes in contemporary journalism, where much column space is devoted to reports about coming events and supposed development trends (cf. Väliverronen 2002). The governmental attitude to the citizen has also changed. Citizens are no longer masses or a group to be taken care of or patronized; rather, they are now regarded as resources, which have a responsibility to educate themselves in order to increase their ability to cope competently with the flows of information in the modern Finnish society (cf. Häyrinen-Alestalo and Pelkonen 2004, 187–8).

The end of benevolent globalization and Finnish energy policy

The aim of this chapter was to analyse the temporary nature of societal risk evaluation. The analysis of the Finnish nuclear decisions in 2001 and 2002 has hopefully convinced the reader of the fickleness of collective risk assessment. National peculiarities such as cultural values and the characteristic features of the national decision-making system are preconditions of societal decision making, but not enough to explain the content of the decisions. Through an analysis of the contextual factors and the broader societal developments it has been shown why the Parliament of Finland made two favourable nuclear decisions. The study pointed out the transformation that the country underwent in the 1990s. The renewal of the country could be seen in the changes in media, lobbying, science, environmentalism and other values, societal structures and national strategy.

However, Finland changed not only of its own accord, but as part of broader international interaction networks. Once more we come full circle to ask what kind of Finland made these favourable nuclear decisions. Eventually, we must forget the cliquish perspective and extend the range of our knowledge. The new Finland is closely connected to different international structures and actor networks. Although the nuclear decisions were made without the grave economic concerns and social insecurities characteristic of the 1990s the factors of uncertainty and instability nevertheless played an important role. New threats have arisen because of the events of September 11, 2001, which triggered a period of international tension. How will the economic development proceed? What is the direction of Russian society and, in a related point, how much should Finnish energy policy depend upon the imports of energy from Russia. Does the 'War against Terror' also mean a new oil crisis, as in the early 1970s? Among other things, such questions were important when the political elite considered the construction of more nuclear power facilities and, more importantly, the shaping of energy policy.

Heavy-handed treatment of a theme of international development heralds the end of globalization. To put it mildly, starry-eyed neo-liberal ideology had its moment, but was replaced by another kind of globalization, which certainly had an influence on the Finnish nuclear decisions. After the mid-1990s Finnish political decision making was grounded in the belief of benevolent globalization softened by the Scandinavian welfare state. The world after the Cold War appeared to be one of peace, harmony, economic equality, humanism, the easing of political tensions and the disappearance of power blocs. The uninhibited market economy, characterized by free and open international interaction, lasted until it encountered many setbacks such as the tottering of the international economy and the reassessment of foreign and security policy. These factors were not present explicitly in the nuclear debate, but they influenced people's attitudes. The fundamental societal developments predisposed people towards another kind of risk decision.

Notes

1. The chapter is a revised version of the previously published article (Litmanen 2004) in a book (Kojo 2004) which deconstructed the background of Finnish nuclear decisions in 2001 and 2002, analysed the national nuclear debate during the period 2000–02 and described the strategies of the different parties involved in nuclear power politics. The first English version of the chapter was a paper presented at the BSA Risk and Society Study Group 'Taking Stock of Risk' Conference, University of Nottingham, 5–7 September 2004.

2. Moral order (what is considered right and wrong, proper and improper) protects the social order of each community. That is to say, it protects the existing social organization. For Douglas, moral order is sanctioned through dangers, fears, and threats, i.e. people acting against the prevailing moral code expect retribution through some sort of a natural punishment (disease, war, witchcraft, divine retribution, crime, natural catastrophe, etc.). Societies and groups do not consider all anxieties equal, but some things are considered more fearsome than others. According to Douglas, societies and groups also choose risks that reflect their social order. Hence, her theory offers the option of explaining fears and anxieties by referring to forms of social organizations. For instance, the fear of pollution, the fear of the depletion of natural resources, or the fear of decreasing biodiversity have become mechanisms through which a society protects its own institutions. In this sense, the anxiety about environmental pollution in an industrial society is functionally equivalent to the fear of ritual pollution in primitive societies. Their function is the same: they protect moral order, which is vital for the preservation of a community's social order (Douglas, 1986, 60; 1992, 5–11; Milton, 1996, 90; Lupton, 1999a, 43–4).
3. In point of fact, in addition to the structuralist emphasis, the Douglas theory also displays a constructionist emphasis. In her theory, as Milton (1996, 96) stresses, public anxiety is produced and sustained not only by objectively existing preconditions, but also by demands expressed by groups and individuals. Her main idea is that knowledge and understanding are constructed in daily interaction between people who are dependent on each other. In a way, risk is a sociological paradox since it can be studied as a collective construction on the one hand, but on the other hand so-called 'objective risks' construct collective consciousness (see, for example, Lupton 1999a). As a rule, sociological risk research has previously focused on the question of how risks are socially constructed and why we accept certain risks and not others. However, risks also construct our collective consciousness, even create new collectives and communities, which declare the eradication of risks, reduction of risks, more equal distribution of risks, or better societal management of risks as their goals.
4. The cultural risk theory of Mary Douglas has its roots in Durkheimian theory, where the most important questions are: How is the existence of a society possible and what holds it together? The answer to both of these questions is 'emotional forces': trust, social bonds, moral obligations, and moral solidarity. These forces, which constitute the cultural and institutional continuity of a society, maintain its structure. Furthermore, it is believed that the structure exists *sui generis*, regardless of individual thought or action. The implicit methodological rules of Durkheimian theory specify in order to understand a society we must look at the macro-level, rather than observing the behaviour of individual actors. This macro-sociological approach involves a high degree of determinism; according to it, most of the actions of an individual can be derived from macro-level forces above him/her (Milton 1996, 89–105).
5. Tommola (2001, 87–90) interviewed those reporters who had made nuclear waste news. Her findings were that journalists regarded the nuclear waste company Posiva as being a more important source of information. Posiva was felt to be dynamic, competent and positive in terms of the distribution of information. Reporters spoke particularly highly of them helping their everyday work in the media. The background material and visits to their own and other countries' similar nuclear waste companies facilitated the execution of their tasks and reduced the amount of background work. The visualization of the news was easier when the reporters could

use the material produced by the enterprise, in the form of graphics, videotapes and television films.

6. According to the dictionary, particular refers to individual things, cases or events as opposed to a general situation or theory. The term particularism has its roots in theology and politics. The theological meaning entails the idea that the existence of God is connected to a single nation or people. The political usage of the term refers to the efforts of a state or people to gain as much independence as possible. By contrast, the term universalism refers to the philosophical idea that there is only a single reality, which forms a totality and from totality we can explain details. The other meaning is that the direction of moral activities is totality. From universality, universal applicability, and broadmindedness, the universities and sciences have moved towards details, particular cases, special needs and individual cases.
7. One common notion in analysing the negative decision was the so-called Chernobyl effect. This is not a sufficient explanation, because opinion polls indicated that at the beginning of the 1980s public opinion in different countries was already critical of nuclear power. If we look for only a single explanation, the serious neglect in nuclear waste management by both of the Great Powers, the Soviet Union and the United States, during the period of the Cold War might be nearer the truth. The neglect was much discussed at the beginning of the 1990s, soon after the collapse of the Soviet Union and the dissolution of the bipolar world order. More historical profundity results if more emphasis is given to the so-called spillover effect. The splashes from the popular anti-nuclear weapon movement reached the public as nuclear power was part of the security secrets and foreign policy of the Cold War period and the world was dominated by the image of global nuclear destruction. Some historians tried to reach even farther in analysing the roots of science phobias and cultural myths behind the fear of radiation.
8. The mobile phone manufacturer Nokia has decisively affected the large-scale change in the production structure. The GNP and export share of the electricity and electronics industry has grown. For instance, the share of exports rose from 11 per cent in 1980 to 30 per cent in 1999. The symbol of this growth has been Nokia, but we must not forget that there have also been other actors, such as Nokia's subcontractors.
9. The government produced a report about the possible future of the country to Parliament in 1997 (VNS 1997). The key vision was summarized in the name of the report 'Reilu ja rohkea - vastuun ja osaamisen Suomi' ('Fair and Courageous – Finland Responsible and Capable'). The report outlines a programme in which Finland would be a European testbed of the information society. The future committee of Parliament shared the view in its report in 1998, but stressed that there are some problems and risks in any such strategy.

References

Bauer, M. (1997) Towards a Functional Analysis of Resistance. In M. Bauer (ed.), *Resistance to New Technology: Nuclear Power, Information Technology and Biotechnology*. Cambridge: Cambridge University Press, pp. 400–10.

Beck, U. (1992) *Risk Society: Towards a New Modernity*. London: Sage Publications.

Beck, U. (1995) *Ecological Politics in an Age of Risk*. Cambridge: Polity Press.

Bertilsson, T.M. (2003) The Social as Trans-Genic. On Bio-Power and its Implications for the Social. *Acta Sociologica* 46(2), 118–31.

Burgess, A. (2004) *Cellular Phones, Public Fears, and a Culture of Precaution*. Cambridge: Cambridge University Press.
Castells, M. and P. Himanen (2002) *The Information Society and The Welfare State: The Finnish Model*. Oxford: Oxford University Press.
Douglas, M. (1986) *Risk: Acceptability According to the Social Sciences*. London: Routledge and Kegan Paul.
Douglas, M. (1992) *Risk and Blame: Essays in Cultural Theory*. London: Routledge.
Douglas, M. and A. Wildavsky (1983) *Risk and Culture: An Essay on the Selection of Technical and Environmental Dangers*. Berkeley: University of California Press.
Eurobarometer 62 (2004) *Public Opinion in the European Union: National Report. Executive Summary. Finland*. (europa.eu.int/comm/public_opinion/archives/eb/eb62/eb62_fi_exec.pdf – accessed 9 August 2005.)
EVA (1999a) *Indicators of Finnish Society*. (http://www.eva.fi/english/julkaisut/indic99/ – accessed 12 September 2002.)
EVA (1999b) *Finnish Attitudes 1999*. (http://www.eva.fi/english/julkaisut/opin99/ – accessed 12 September 2002.)
EVA (2001) *EVA-raportti suomalaisten asenteista 2001* (EVA-report on Finns' Attitudes 2001) (http://www.eva.fi/julkaisut/asenne2001/ – accessed 12 September 2002.)
Forsander, A. (ed.) (2002) *Immigration and Economy in the Globalization Process: The Case of Finland*. Sitra Reports Series 20. Helsinki: Sitra.
Gibbons, M., C. Limoges, H. Nowotny, S. Schwartzman, P. Scott and M. Trow (1994) *The New Production of Knowledge: The Dynamics of Science and Research in Contemporary Societies*. London: Sage Publications.
Hautamäki, A. (2001) *Finland in the Face of Change: A Report on Finland's Challenges*. Sitra reports series 19, Helsinki: Sitra.
Häyrinen-Alestalo, M. and A. Pelkonen (2004) Teknologiapolitiikan näkymättömät kansalaiset. (Invisible Citizens of Technology Policy). In T. Lemola and P. Honkanen (eds), *Innovaatiopolitiikka. Kenen hyväksi, keiden ehdoilla?* (Innovation Policy. For the good of who, for which prerequisites?). Helsinki: Gaudeamus, pp. 177–93.
Hylkilä, I. (2003) *Suomen viidennen ydinvoimalan lobbaus kansanedustajan näkökulmasta* (The lobbying of fifth nuclear power plant in Finland from the perspective of the Member of Parliament). University of Jyväskylä, Masters thesis in Organizational Communication & PR. (http://selene.lib.jyu.fi:8080/gradu/v04/G0000435.pdf – accessed 13 February 2004.)
Ilmonen, K. and K. Jokinen (2002) *Luottamus modernissa maailmassa* (Trust in the modern world). Jyväskylä: SoPhi.
Kerkelä, H. (1996) *Vanhan maailman peilissä. Modernin yhteiskunnan synty ja pohjoinen aineisto* (In the mirror of the old world: The birth of modern society and Nothern Data). Helsinki: Gaudeamus.
Kojo, M. (ed.) (2004) *Ydinvoima, valta ja vastarinta* (Nuclear power, power and resistance). Helsinki: Like.
Konttinen, E. (1991) *Perinteisesti moderniin. Professioiden yhteiskunnallinen synty Suomessa* (Traditionally to modernity: The societal birth of professions in Finland). Tampere: Vastapaino.
Kyllönen, S. (2004) Ydinvoiman ilmastomuutos (The climate change of nuclear power). In M. Kojo (ed.), *Ydinvoima, valta ja vastarinta* (Nuclear power, power and resistance). Helsinki: Like, pp. 51–86.
Lammi, H. (2004) Tarinat kovasta ytimestä (The stories concerning the hard nucleus). In M. Kojo (ed.), *Ydinvoima, valta ja vastarinta* (Nuclear power, power and resistance). Helsinki: Like, pp. 11–50.

Lidskog, R. and T. Litmanen (1997) The Social Shaping of Radwaste Management: the Cases of Sweden and Finland. *Current Sociology* 45(3), 59–79.

Litmanen, T. (1996) Environmental Conflict as a Social Construction: Nuclear Waste Conflicts in Finland. *Society and Natural Resources* 9(5), 523–35.

Litmanen, T. (2001) *The Struggle over Risk: The Spatial, Temporal, and Cultural Dimensions of Protest Against Nuclear Technology*. University of Jyväskylä, Jyväskylä Studies in Education, Psychology and Social Research, No. 177.

Litmanen, T. (2004) Suomen ydinvoimaihme (Finnish nuclear miracle). In M. Kojo (ed.), *Ydinvoima, valta ja vastarinta* (Nuclear power, power and resistance). Helsinki: Like, pp. 209–30.

Lupton, D. (1999a) *Risk*. London: Routledge.

Lupton, D (ed.) (1999b) *Risk and Sociocultural Theory: New Directions and Perspectives*. Cambridge: Cambridge University Press.

Maula, M.V.J., J.-P. Salmenkaita and M.A. Uusitalo (2004) Teknologia- ja innovaatiopolitiikan perusteet ja perushaasteet (The foundations of technology and innovation policy and its basic challenges). In T. Lemola and P. Honkanen (eds), *Innovaatiopolitiikka. Kenen hyväksi, keiden ehdoilla?* (Innovation Policy: For the good of who, for which prerequisites?) Helsinki: Gaudeamus, pp. 281–9.

Milton, K. (1996) *Environmentalism and Cultural Theory: Exploring the Role of Anthropology in Environmental Discourse*. London: Routledge.

Nieminen, H., P. Saarikoski and J. Suominen (1999) Uusi media ja arkipäivä. Johdannoksi (New media and everyday. Introduction). In H. Nieminen, P. Saarikoski and J. Suominen (eds), *Uusi media ja arkielämä. Kirjoituksia uuden ajan kulttuurista* (New media and everyday: Writings on the culture of new era). University of Turku, School of Art, Literature and Music, Media Studies. Publications, Series A, no. 41, 1–7.

Nieminen, M. and E. Kaukonen (2001) *Universities and R&D Networking in a Knowledge-based Economy: A Glance at Finnish Developments*. Sitra Reports Series 11. Helsinki: Sitra.

Nissinen, J. (2004) Valtavirran vastarinta (The resistance of Valtavirta). In M. Kojo (ed.), *Ydinvoima, valta ja vastarinta* (Nuclear power, power and resistance). Helsinki: Like, pp. 87–105.

Nowotny, H., P. Scott and M. Gibbons (2002). *Re-thinking Science: Knowledge and the Public in an Age of Uncertainty*, reprinted. Cambridge: Polity Press in association with Blackwell Publishers.

Pesonen, P. and O. Riihinen (2002) *Dynamic Finland: The Political System and the Welfare State*. Studia Fennica, Historica; 3. Helsinki: Finnish Literature Society.

Raittila, P. (2002) Journalismin rooli ydinjätekeskustelussa. Näkökulmien välittäjä vai vuorovaikutuksen organisoija (The role of journalism in the debate of nuclear waste. Supplier of perspectives or organiser of interaction). In P. Raittila, P. Hokkanen, M. Kojo and T. Litmanen (eds), *Ydinjäteihme suomalaisittain* (Nuclear waste miracle in a Finnish way). Tampere: Tampere University Press, pp. 67–91.

Rasilainen, K. (ed.) (2002) *Nuclear Waste Management in Finland: Final Report of Public Sector's Research Programme JYT2001 (1997–2001)*. Ministry of Trade and Industry Finland. Studies and Reports 15/2002.

Rosenberg, T. (1999) Turhauttavaa teatteria. Loppusijoitus-YVA Loviisa-liikkeen näkökulmasta (Frustrating theather. EIA of the final disposal from the perspective of Loviisa movement). In T. Litmanen, P. Hokkanen and M. Kojo (eds), *Ydinjäte käsissämme. Suomen ydinjätehuolto ja suomalainen yhteiskunta* (Nuclear waste in our hands. Finnish nuclear waste management and Finnish society). Jyväskylä: SoPhi, pp. 266–82.

Rosenberg, T. (2004) Kun Loviisa atomisuukon sai (When Loviisa got nuclear kiss). In M. Kojo (ed.), Ydinvoima, valta ja vastarinta (Nuclear power, power and resistance). Helsinki: Like, pp. 159–78.

Ruostetsaari, I. (2003) *Valta muutoksessa* (Power in change). Helsinki: WSOY.

Ruostetsaari, I. (1998) *Energiapolitiikka käännekohdassa. Järjestöt ja yritykset vaikuttajina vapautuvilla energimarkkinoilla* (Energy policy in the turning point. Organisations and enterprises as agents in freeing energy markets). University of Tampere, Department of Political Science and International Relations. Publications 8/1998.

Schulz, W. (2004) Reconstructing Mediatization as an Analytical Concept. *European Journal of Communication* 19(1), 87–101.

Stehr, N. (1994) *Knowledge Societies*. London: Sage Publications.

Stehr, N. (2001) *The Fragility of Modern Societies: Knowledge and Risk in the Information Age*. London: Sage Publications.

Tammilehto, O. (1996) Miksi viidettä ydinvoimalaa ei rakennettu? Ydinvoiman vastaisen liikkeen historiaa (Why was the fifth nuclear power plant not constructed? The history of the anti-nuclear movement). In T. Heima-Tirkkonen, T. Kallio-Tamminen and T. Selin (eds), *Ruohonjuurista elämänpuuksi. Suomalainen vaihtoehtoliikehdintä* (From grassroots to a tree of life. Finnish alternative movements). Vihreä Sivistysliitto, Kosmos-series, no. 3, Helsinki, pp. 97–111.

Tiedebarometri (2001) *Tutkimus suomalaisten suhtautumisesta tieteeseen ja tieteellisteknisеen kehitykseen.* (A study of how Finns perceive science and scientific-technological development). Tieteen tiedotus ry, Yhdyskuntatutkimus Oy. (http://www.saunalahti.fi/~pena/tb/tiedebaro.htm – accessed 9 August 2005.)

Tiedebarometri (2004) *Tutkimus suomalaisten suhtautumisesta tieteeseen ja tieteellisteknisеen kehitykseen* (A study of how Finns perceive science and scientific-technological development). Tieteen tiedotus ry, Yhdyskuntatutkimus Oy. (http://www.sci.fi/~yhdys/tb2/tiedebaro2.htm – accessed 9 August 2005.)

Tommola, A. (2001) Toimittajat ja ydinjäte. Kahdeksan toimittajan käsityksiä suomalaisesta ydinjätekeskustelusta ja sen osapuolista (The journalists and nuclear waste. The views of eight journalists on nuclear waste debate and its parties). In P. Raittila (ed.), *Mediat ydinjätettä hautaamassa. Eri intressiryhmien julkisuuteen pääsy, dialogi ja argumentointi ydinjätteen loppusijoitusta koskevassa keskustelussa 1999–2001.* (The media in burying the nuclear waste. Different stakeholders access to publicity, dialogue and argumentation in the case of final disposal of nuclear waste from 1999 to 2001). University of Tampere. Department of Journalism and Mass Communication, Journalism Research and Development Centre. Publications C 34/2001, pp. 81–100.

VNS (1997) *Reilu ja rohkea – vastuun ja osaamisen Suomi.* (Fair and Courageous – Finland Responsible and Capable.) Report by the committee for the future 1/1998, vp. TuVM 1/1998 vp, VNS 3/1997 vp. (http://www.eduskunta.fi/triphome/bin/thw/?${APPL}=utppdf&${BASE}=utppdf&${THWIDS}=0.3/59351&${TRIPPIFE}=PDF.pdf – accessed 17 March 2008.)

Väliverronen, E. (2002) *Geenin valta. Modernin bioteknologian kansalliset kertomukset ja yleisöt* (The power of genes. National tales of modern biotechnology) University of Helsinki. Department of Communication. (http://www.valt.helsinki.fi/staff/valiverr/geva.ka1b.htm – accessed 16 February 2004.)

Vuori, S. (ed.) (1997) *Publicly Administrated Nuclear Waste Management Research Programme 1994–1996*. Final Report. Ministry of Trade and Industry Finland. Studies and Reports 22/1997.

Part V
Conclusions

8
The Revival of Nuclear Power in a Strong Administrative State

Matti Kojo

The new millennium began with two major nuclear decisions by the Parliament of Finland. In May 2001 the Parliament ratified the Decision-in-Principle (DiP) concerning the final disposal repository for spent nuclear fuel (SNF). One year later, in May 2002, it voted for the construction of a new nuclear power plant (NPP) unit. After the favourable voting result the Parliament decided unanimously to approve the extension of the planned disposal repository. The decisions were unique. At the beginning of the millennium, the difference from other countries was striking because at the same time SNF management in other countries – with the exception of Sweden – had either come to an end or else important decisions were being postponed as the result of political disagreements. Furthermore, some countries were shutting down their NPPs because nuclear power was no longer being viewed as an acceptable source of energy. Today, however, as we move towards the end of the first decade of the new millennium, the situation is already somewhat changed, with several countries reconsidering their nuclear power policies. According to Frogatt (2007a; see also 2007b), the position of nuclear power in the European Union (EU) countries can be classified into the following four categories: (1) countries with an active construction programme (Bulgaria, Finland and France); (2) countries that are to take key decisions in the near future (Lithuania, Romania, Slovakia and the UK); (3) countries which are maintaining the status quo and extending the lives of their current plants (the Czech Republic, Hungary, the Netherlands and Slovenia); and (4) countries which have agreements to phase out their current NPPs (Belgium, Germany and Spain). In addition to Frogatt's list it should be highlighted that Estonia and Poland have discussed a joint NPP project with Lithuania. Estonia has also taken some interest in the Finnish government's plans for the construction of further NPPs in Finland. In February 2009 the Swedish government announced an agreement allowing for the replacement of existing NPPs. In the EU in 2004, 12 out of the current 27 Member States produced no nuclear energy (World Energy Council 2007, 9–10). Finland is the only country that has decided both to continue the construction of NPPs and also to store the

SNF in its bedrock. The Finnish situation is illustrated by the conclusion drawn in the World Energy Council Report (2007, 34) that over the past 25 years Finland has witnessed a complete shift in respect of the public opinions and perceptions of nuclear power. Nevertheless, the possible revival of nuclear power should be seen as a Western European and North American phenomenon because of the developments in the nuclear programmes in Japan, Korea, China and India.

The questions to be considered at the end of this study are: Why and how did this shift take place in Finland? How do the authors explain the recent Finnish nuclear policy decisions? What are the main findings of our chapters? The aim of this final chapter is to summarize the conclusions drawn by the authors. The focus will be on understanding political decision making in relation to the further construction of nuclear power in Finland. Accordingly, the structure of the chapter is as follows. First there is an introduction of the decision-making process, the DiP process concerning nuclear facilities. The chapter then takes a closer look at the character of actual decision making by showing and explaining some informal features of Finnish decision making. At the theoretical level the focus is on the idea of a strong administrative state presented in Chapter 5 by Säynässalo and at the practical level on outcomes of the partnership between the administration and the nuclear utilities presented in Chapter 2 by Lampinen. Following this, there is a discussion of the changes in the pro- and anti-nuclear lobbies in Finland following the parliamentary rejection of 1993. The observations on lobbying are based on the contribution made by Lammi in Chapter 3. Nuclear waste management and the latest disagreements inside the Finnish nuclear industry camp are handled as a section of their own – based on the contribution of Kojo (Chapter 6). The chapter ends with the view of discourses, knowledge production and risk perception around nuclear policy decision making in Finland. Chapter 4 by Berg and Chapter 7 by Litmanen help to outline the mental landscape of societal decision making in Finland.

Formal decision making on nuclear facilities in Finland

In total, the formal decision-making process in respect of the new NPP unit took almost four years, from the beginning of the EIA process in June 1998 to the parliamentary ratification of the DiP for the new NPP unit in May 2002 (see Appendix 1 at the end of this chapter). It was a further three years before the issuing of the construction licence by the Council of State in February 2005. The description of the formal decision-making procedure offers a smoother picture of the process in Finland than is actually followed in the actual course of events. However, the authors of the book highlight some informal processes which compel us to re-evaluate the democratic aspects of Finnish decision making. Before studying the informal processes more closely, the formal decision-making process is introduced in detail.

The formal decision-making procedure included in the Nuclear Energy Act is termed the Decision-in-Principle (DiP) process. This process is launched when a company applies for the DiP concerning a nuclear facility from the Council of State, as did Posiva in May 1999 and TVO in November 2000 (Appendixes 1 and 2).[1] Thus, the nuclear industry acts as a key initiator of policy formation and implementation as argued by Säynässalo in her chapter, in which she explains the basis of the legal-administrative framework of the licencing procedure. However, it should be taken into account that the utilities tried to secure advance political guarantees for their future business activities. The debate – or, more precisely, the public demands on reforming the government's steering ideology on the nuclear energy policy – has taken place since the mid-1990s. Pro-nuclear voices insisted that the process should be conducted on very different lines to those that had been adopted during the process in the early 1990s when the pro-nuclear lobby was unsuccessful and the Finnish Parliament rejected the DiP. The pro-nuclear lobby proposed that the government should develop a long-term policy on nuclear energy policy before the submission of the application. This call was answered by the then minister in charge of the Ministry of Trade and Industry, Antti Kalliomäki (of the Social Democratic Party). After the ratification on the new NPP unit in May 2002 similar demands were made by PVO, the main shareholder in the nuclear utility TVO. On this occasion the pro-nuclear lobby argued for continuation of the policy that had already been accepted in 2002.

In Finland the Ministry of Trade and Industry[2] is the authority responsible for the preparation of the DiP. The Group of Energy Governance and Nuclear Energy of the Energy Department of the MTI prepares the decisions of the ministry and the Council of State. According to the Nuclear Energy Act, in the processing of the issues the ministry is advised by the Advisory Committee on Nuclear Energy, whose membership consists of experts in the various areas.[3] The specified tasks of the committee are: to monitor the development of the nuclear energy sector and to present motions and suggestions for steps necessary in the sector; to issue statements; to make suggestions concerning the development and organization of research and training related to the nuclear energy sector; to maintain connections with the Advisory Board for Nuclear Safety and the Advisory Board for Radiation Safety; and to handle other issues assigned to the committee by the government or the Ministry of Trade and Industry.

The environmental impact statement (EIS) of the nuclear facility is appended to the DiP application. Thus, the DiP process is preceded by the environmental impact assessment (EIA) process, including participatory initiatives organized by the developer (see, for example, Hokkanen and Kojo 2000). The written opinions requested by the contact authority are the minimum level of public participation. According to the EIA Act, the MTI acts as the contact authority responsible in relation to any plans concerning nuclear facilities. In all other types of EIA plans, the contact authorities are Regional

Environmental Centres under the authority of the Ministry of Environment. The information needed for the DiP application is stipulated under the terms of the Nuclear Energy Act and Decree.

According to the Nuclear Energy Act, the MTI is also responsible for asking for compulsory statements from actors such as the Finnish Radiation and Nuclear Safety Authority (STUK), some ministries, and the municipality of location and neighbouring municipalities in the DiP process. STUK, as the radiation and nuclear safety authority, is responsible for the delivery of the preliminary safety statement, including the statement of the advisory committee to the MTI. According to the Nuclear Energy Act, the MTI is also responsible for arranging a public meeting in the municipality that is the proposed site for the new nuclear facility. According to the Act, the focus is on the residents of the candidate municipality and the neighbouring municipalities. In practice, anybody affected by the facility is allowed to give his or her opinion to the meeting, either orally or in writing. Written opinions can also be submitted direct to the ministry. However, no debate or deliberation is allowed in the public meetings, which are arranged only in the proposed municipality. A meeting consists solely of statements of the affected and the only opportunity to make a response is following the public meeting, when the MTI delivers all of the feedback to the applicant, who is then able to reply to the comments. The MTI also submits a summary of the statements and opinions delivered to both the government and the applicant.

The Nuclear Energy Act includes a right of veto for the council of the municipality in which territory the proposed facility would be located. STUK is also vested with a right of veto over safety issues. Thus, the municipal council has the power to prevent the location of the nuclear facility into the municipality, which creates a local balance of power between the municipality and the applicant. The power of the municipal council is based on the potential use of veto, whereas the applicant has the resources to conduct research and information activities. Therefore, in order to build local support for their schemes the Finnish nuclear industry has, for example, established cooperation groups with the political representatives of the nuclear communities (see Chapter 6 of this volume; Kojo 2005).

According to Miettinen and Väliverronen (1999, 16), the Finnish orientation is clearly towards the specialist-driven assessment of technology. This orientation can be seen as emphasizing the fulfilment of the information needs of decision making. In 1995 the task group appointed by Parliament to prepare a proposal for organizing technology assessment within the Parliament was not interested in either the Danish or the Dutch experiences, where the orientation of technology assessment was focused more on the extended public debate on technology. The initiator was MP Martti Tiuri (National Coalition Party), a professor of radio technology and the chair of the Futures Committee of Parliament. Professor Tiuri is familiar in Finland as a renowned advocate of nuclear power. According to Miettinen and Väliverronen

(1999, 16), the preparatory group of Parliament was not interested in the ideas of participation, democracy or in the involvement of stakeholders in technology assessments. For example, a consensus conference was not regarded as a serious form of technology assessment. As early as the late 1980s Pentti Malaska and Pirkko Kasanen (1987, 56–9) suggested that in order to enhance the decision making and the public debate on nuclear energy a technology panel should be established, a referendum should be held and there should also be an expansion of the value base of STUK's top management. The studies of the citizens' involvement practices of the Finnish nuclear sector (Hokkanen and Kojo 2000; Hokkanen 2007) indicate that the approach adopted by the Finnish nuclear industry does not include deliberative features, although some development can be seen when it is compared to the practices that had been applied in the 1980s and early 1990s (Kojo 2005).

The MTI submits the DiP application to the government after it has studied the application and the statements and opinions provided by different stakeholders. The government then formulates the DiP, in which it considers the overall benefits to society. In accordance with the Nuclear Energy Act, particular attention is paid to the need for the nuclear facility project with respect to the country's energy supply, the suitability of the intended site for the nuclear facility, its effects on the environment and the arrangements for the management of nuclear fuel and nuclear waste. Following this process, the DiP still has to be ratified by Parliament. According to the Nuclear Energy Act, Parliament is unable to change the content of the decision. However, the application can be changed as the result of political pressure on the part of either the government or an individual political party. This was the case, for example, with the DiP application of Posiva because the application was reformulated by the applicant in November 2000. After the TVO's application on 15 November the parliamentary group of Greens and their Minister of the Environment, Satu Hassi, made it clear that the Greens were not prepared to approve the applied capacity of the facility because they wanted to avoid any connection being made between the policy decisions relating to nuclear waste disposal and the construction of further NPPs. The Greens did not want to give a positive signal for the construction of a new NPP unit by approving Posiva's original application.[4] The disposal capacity included in the application was reduced by Posiva in a letter sent to the Ministry on 23 November 2000 to cover the SNF produced in the four existing units. The original application submitted on 26 May 1999 was dimensioned for SNF produced in the six NPP units. Thus, the amount was decreased from the maximum 9,000 tU for which Posiva had originally applied to just 4,000 tU. Posiva asked the Council of State to make a decision about the disposal of SNF produced in the new NPP unit, 2,500 tU, at the same time as it decided about the DiP concerning TVO's application. Furthermore, Posiva stated that its application related only to the SNF produced in the one new NPP unit (Posiva 1999, 3, 7; Council of State 2000). The SNF which would be produced in the sixth unit, 2,500 tU, was excluded in order to avoid speculation over the construction of additional NPPs and to mitigate fears of SNF imports.[5] The decreased amount can be also seen as a concession to the opponents of the disposal plan. The Minister of the Environment, Hassi (2002, 95) (of the Green League), also tried to postpone the schedule of the DiP, but this initiative was not supported in the government.

The parliamentary process of DiP also includes an expert hearing procedure in the committees. In the case of the fifth NPP unit nearly 200 experts were heard in eight different parliamentary committees. These committees were perceived by MPs as important background groups and also as privileged opportunities to obtain knowledge (see Chapters 2 and 4; on the parliamentary process regarding the final disposal facility, see Raittila and Suominen 2002, 92–4). All of the committee meetings are held behind closed doors, thus it is impossible to offer any evaluation of the discussions that have taken place between the individual committee members.

According to the report issued by the World Energy Council (2007, 72–5), the main components of the Finnish approach were that: (1) the processes, both the EIA and the political approval process (DiP), were performed in advance of the industrial decision process. The advantage of the EIA was that it involved the decoupling of data in order to ensure its validity independent of different reactor designs and to provide information in support of the political approval process. The report pointed out that the Finnish Nuclear Energy Act states that a DiP is required before the industrial project can be started. The purpose is to obviate political interference with the regulatory process, once the DiP is gained. The report concluded that the DiP is thus the final step in the political decision-making process and authorizes TVO to continue preparations on both the commercial and technical levels for the construction of a new NPP. In addition to these, (2) the feasibility study of the 'candidate designs' of the NPP unit and (3) a well-defined regulatory context, were mentioned as being among the main components of the Finnish approach.

The Ministry of Trade and Industry at the core of administration-led decision making

According to Lampinen, the MTI has managed to establish an excellent public–private partnership with the applicant companies which could guarantee the smooth handling of their applications regardless of public criticism. The development of the partnership can be explained, at least in part, because of the similar educational backgrounds and shared career paths of the relatively small number of experts in the nuclear sector. Thus, MTI is seen as the principal gatekeeper in the field of Finnish energy policy. MTI's strong role is based on its power to influence – or even to guide – the preparatory processes, which is safeguarded in turn through the control of energy research policy and funding in Finland.

The public–private partnership covered the Finnish utilities' relationship not only with the MTI but also with STUK. According to the report of the World Energy Council (2007, 83), the instruction phase for the delivery of the construction licence of the NPP unit was relatively quick (14 months; see Appendix 1), because for many years the feasibility studies of 'candidate designs' were carried out by the utilities and STUK. The aim was to ensure the absence of those safety issues that were preventing compliance with Finnish nuclear safety regulations. The study covered most of the alternative plants presented in the DiP application of TVO and thus the gathering of information in advance made it possible to make the statement quickly, following the launch of the formal licensing process. Such a study was proposed, for example, as a part of the NPP safety research in 1998. The Advisory Committee on Nuclear Energy (1998, 10) stated that the publicly funded research programme should also maintain the capability for research and development needs in the case of new power plant projects. The level of public funding for nuclear safety and nuclear waste research increased in Finland following the Chernobyl accident in 1986 and also in the early 1990s due to the discussion of the further construction plans. Following the rejection of the NPP proposal in 1993 and in part as a result of the economic downturn public funding was reduced over the course of the mid-1990s. Since 1996 funding has again increased as the result of Finland's involvement in EU research programmes. In 1996 the volume of funding was €30.1 million, with more than two-thirds of this sum being covered by the nuclear utilities (Advisory Committee on Nuclear Energy 1998, 10–11). The use of public research programmes has been found to yield results in terms of maintaining and creating know-how in the nuclear sector (MTI 2000).

In his chapter Lampinen analyses the arguments used in the DiP process for the justification of the new NPP unit. He states that, in principle, the Nuclear Energy Act offers a very broad and democratic decision-making process. However, despite the fact that it has in principle a democratic starting point, in practice power is concentrated in the hands of a few civil servants at the Energy Department of MTI.[6] In fact, one active anti-nuclear lobbyist has identified the Head of the Energy Department as the leading pro-nuclear lobbyist (Klötzer 2006, 23). According to Lampinen, the MTI is in overall charge of Finnish energy policy and acts, either by itself or through the actions of VTT, as the main expert in energy policy, climate policy and energy technology-related affairs both within the government and in Parliament. Furthermore, it also controls energy research policy and funding in Finland, either directly or through the operations of VTT and TEKES.

Lampinen argues that one result of the concentration of power has been MTI's ability to prevent external expert arguments from being given serious consideration. Even the Ministry of the Environment has had great difficulties in influencing the preparation process (Hassi 2002, 313–15, 317–19). For example, during the nuclear debate at the beginning of the millennium, the MTI was reluctant to prepare a scenario consisting of a renewable energy source and a reduction in the use of energy (Lammi 2004, 32). Lampinen demonstrates the pitfalls of the Finnish review process. His analysis of the arguments stated in the DiP application of the fifth unit as a justification for the overall good of society indicates that some of TVO's arguments are valid, whereas others are unlikely to be correct and some have already been shown to be false. Among the latter are the current minimum construction cost estimates, which are now thought to be as much as 100 per cent higher than the upper limit of the cost range given in TVO's application. In addition, the current estimate for reactor construction time is at least 70 per cent longer than the estimate submitted in the application, a delay which will also have significant cost implications. Lampinen notes how the major arguments for getting support from the Finnish MPs included the relatively low costs and rapid building time and also its impact in terms of fulfilling the Kyoto Protocol targets. The malfunction of the review process and the role of the MTI as its gatekeeper raises the question of how the review process itself could be reformed and thus made more transparent and reliable. As a result, the capability and the motives of MTI to conduce a reliable review of the application and the related comments are called into question. Furthermore, several fundamental aspects of the decision-making process that are assessed by Grimston (2005) as being strengths of the Finnish model – such as a realistic timescale, fairness and transparency in the decision-making process and independent review – appear on closer examination to be invalid.

Lampinen concludes that the MTI has been able to create a hegemonic political discourse in which only a very narrow range of alternatives to their choice are actually relevant, although from the outside it may seem that a much wider set of alternatives is under consideration. This conclusion is reinforced in Chapter 3, where Lammi shows that the MTI has repeatedly ignored parliamentary and NGO calls to produce a third scenario. Lampinen argues that the MTI has held the view that energy policy should be market driven, whereas EU-level policy emphasizes an active government role in renewable energy and the promotion of energy conservation. For example, discussion about the preparation of the RES-E directive promoting electricity from renewable energy sources is one example which shows, according to Lampinen, that the MTI protected the building of the new NPP unit. The MTI convinced the Parliament of Finland that the target level set by the EU Commission and the EU Parliament would be impossible to achieve and that no binding targets should be accepted.

Chapter 5 by Säynässalo offers another view of the strong role played by administration in the development of Finnish nuclear energy policy. In contrast to Lampinen's pragmatic approach, Säynässalo introduces the theoretical framework of a strong party state and a strong administrative state in order to explain nuclear energy policy in Finland and to make some comparisons with other countries. She contrasts the ability of Finland,

France and Sweden to assert control, through their constitutional structures and administrative machinery, over nuclear energy policy. Säynässalo argues that somewhat similar institutional dynamics lie behind the nuclear policy processes in France and Finland – two countries that have, in recent years, been characterized as pro-nuclear. Although the Finnish Parliament is the supreme authority in decision making, the administrative branch of the government has been able to lay down the basic rules of the parliamentary process. Säynässalo argues that in Finland the relationship between the state and the nuclear utilities is characterized by a legal-administrative alliance, as in France. Lampinen draws the same kind of conclusion, in the form of a public–private partnership, at the level of practice. According to Säynässalo, the Nuclear Energy Act of 1987 paid a greater degree of attention to the political aspect than had been the case in the earlier Atomic Energy Act of 1957. However, the new Act did not contribute to any alteration of the basic legal-administrative alliance between the state and the nuclear utilities. At this stage the process for making decisions on nuclear energy was still in its nature of a licensing procedure. Similarly, even though Parliament did reject the construction plan in 1993, they did not manage to call into question the core of the alliance, namely the licensing procedure, and the legal-administrative framework for a decision-making procedure (see Chapter 5). The alliance stayed non-political in nature even at the beginning of the twenty-first century under the pressure of the debate on climate change. The state, and the new government of Prime Minister Matti Vanhanen (Centre Party), gave only weak signals in favour of the construction of further NPPs, although the three main parties – the Social Democratic Party, the Centre Party and the National Coalition Party – seemed to agree about the need for a sixth NPP unit before the parliamentary elections of 2007. The guiding ideology of Vanhanen's government is stated in the governmental programme in which the government's nuclear energy policy and position in relation to the submission of a DiP application is made clear by the fact that it did not exclude the nuclear option as a form of energy production.

Säynässalo states that the Finnish state has lacked a substantial national programme in support of the use of nuclear energy. In Finland, the state has adopted the role of a regulator, which is in contrast as the active and promotional role adopted by the state in both Sweden and France, where party governments decided to pursue nuclear energy programmes soon after the oil crises of 1973 and 1974. It should be pointed out, however, that in the early 1970s a massive further development of the nuclear energy programme was planned in Finland, with the intention of increasing the electricity production capacity from 4,000 MW in 1970 up to 10,000 MW by 1980. This would have meant the construction of several new NPP units (Energy Policy Committee 1972, 8, 27; Sunell 2001, 79). However, this plan was never accepted by the government. Säynässalo stresses that the state-owned IVO negotiated with the Soviet partner (Atomenergoeksport) over the contract of the third unit of Loviisa during the 1970s, albeit in the absence of overt government approval. Likewise, instead of struggling over a national nuclear programme, the Finns began to struggle over the fifth unit since the late 1970s (on the anti-nuclear movement see Tammilehto 1994). As the result of the Three Mile Island accident in 1979 the acquisition contract of the third unit of Loviisa between IVO and the Soviet Atomenergoeksport was never signed (Michelsen and Särkikoski 2005, 256–7).

According to Säynässalo, the fact that the Finnish government never made a decision to establish a coherent national nuclear programme shows why the decision-making procedure set by the Nuclear Energy Act has been generally accepted by the anti-nuclear lobby as being a 'neutral' framework for making decisions in respect of nuclear energy. No party government has ever tied its hands on the nuclear issue. In addition, the administration has not supported nuclear power as openly as has been the case in France (Hecht 1998). Although nuclear energy in Finland as a technology process is characterized by strong nationalistic features and at the national level by a strong establishment, this marks a considerable difference between France and Finland.

For Säynässalo, the similarities between the institutional dynamics of nuclear energy decision making in France and Finland culminate in the fact that politicians and parties opposed to nuclear energy have been incapable of resisting the strong independent power of the administration. The independent powers of the administration have constituted just such a state mechanism that has pre-empted the attempts of reform-oriented parties to defend and promote critical alternatives. This, in turn, has made the political parties in both countries consensus-oriented with regard to the rules of the game laid down by administrative bodies; as a result, they are weak channels for the advancement of anti-nuclear arguments. It appears to be common to the compromised position of political parties in both of these countries that their political culture is inclined to stress the nature of political parties as self-interested actors, whereas the administration appears to be a more plausible promoter of the interests of the whole nation.

Thus, in the present Finnish system the MTI is a gatekeeper that has no strong political guardian. The phasing out of nuclear power has never been a major issue on the Finnish political agenda, although some minor parties, such as the Green League, the Swedish People's Party and the Left Alliance, have included the issue of the nuclear phase-out in their political manifestoes. However, a number of governments have refused to advocate the construction of NPPs. For example, following the Chernobyl disaster in 1986 Prime Minister Harri Holkeri, of the National Coalition Party government, which was in office from 1987 to 1991, refused to discuss the issue of further construction. The Finnish Parliament's decision to reverse the government's DiP in September 1993 also had an impact on Finnish nuclear energy policy; in the period 1995–99 Prime Minister Paavo Lipponen's government did not place the further construction of NPPs on the political agenda, although some

MPs planned proposing a resolution in favour of further construction (see Hylkilä 2003, 68–9; Elo 2007, 103). However, the energy strategy prepared by Lipponen's government in 1997 no longer excluded nuclear power as a form of energy production and, later, the strategy was adopted as part of the governmental programme of Prime Minister Lipponen's second government (which was in office in the period 1999–2003). The decision not to exclude the further construction of NPPs was interpreted by the Finnish nuclear energy industry as a green light for them to apply for the DiP (Kyllönen 2004, 56–7). In summary, it can be suggested that there has never been strong political pressure in favour of the phasing out of NPP units in Finland. The researcher Milka Sunell has pointed out ironically that even the Finnish anti-nuclear movement has always only been against the further expansion of nuclear power (*Helsingin Sanomat* 30 April 2006).

Changing styles of the pro-nuclear lobby and the anti-nuclear movement

One important area to consider in the Finnish nuclear policy decision-making process is the role of lobbying. In the biography of Tauno Matomäki,[7] Antti Tuuri (2005, 276) wrote that the lobbying project in respect of a favourable decision on the new NPP unit had been, alongside its campaign to secure EU membership, the largest ever undertaken by the Confederation of Finnish Industry and Employers. Unfortunately, there is currently very little information available in relation to the lobbying activities and strategies of the Finnish pro-nuclear parties in the case of nuclear energy policy. For example, the history of the Confederation of Finnish Industry and Employers (Mansner 2007) largely excludes the subject of pro-nuclear lobbying. There is a common impression that in the early 1990s pro-nuclear lobbying failed partly because of the bluntness of the strategies adopted in cabinet (Ruostetsaari 1998, 158–60; Hylkilä 2003, 76). Matomäki's biography makes no direct admission of this kind of mistake. However, a learning process was launched because of their failure in the planning process of 1993, which was thought to have failed largely because of the overuse of jargon-riddled technical briefings, which were offputting to the public. The pro-nuclear lobbyists of the industry assumed that the public was more attracted by the emotional appeals from the Greens (Tuuri 2005, 271–2; see also Hylkilä 2003, 69–70).

According to Matomäki's biography, in 1997 a new struggle over the fifth reactor began, inspired by the signing of the Kyoto Protocol. However, according to Hylkilä (2003, 68–9, 76), the basis for the nuclear issue had already been prepared in 1995–96 when a group of MPs planned to propose a resolution. According to the former MP Mikko Elo (Social Democratic Party), the MPs Martti Tiuri (National Coalition Party), Sirkka-Liisa Anttila (Centre Party) and Elo himself were all active in the pro-nuclear lobby following the elections of 1995. In addition to Elo, two Social Democratic Party Group

MPs, Reijo Laitinen and Matti Puhakka, were also lobbying in favour of nuclear power. Prime Minister Lipponen, however, was careful because of the new government coalition, which also included the Green League (Elo 2007, 103–4). Lipponen's wide-ranging rainbow coalition intended to steer the Finnish state out of its economic difficulties following the deep depression of the early 1990s.

At the end of 1998 a new group of four MPs – Matti Huutola (Left Alliance), Timo Kalli (Centre Party), Reijo Laitinen (Social Democratic Party) and Juhani Sjöblom (National Coalition Party) – established a pro-nuclear lobby (Hylkilä 2003, 76–97; Yle 2005). Since this time the group has met on a regular basis and it has also grown in size – by 2002 the lobby group consisted of around twenty MPs. TVO and other sections of the industry also contacted the MPs. Two pro-nuclear working parties were established: one within the Confederation of Finnish Industry and Employers and the other within the nuclear power company TVO, representing the owners of the company. Both of the working parties were chaired by Tauno Matomäki and they held meetings over a period of nearly four years. The industry cooperated with the lobby group established by the Central Organization for Finnish Trade Unions (SAK). The Central Union of Agricultural Producers and Forest Owners (MTK) and the Chamber of Commerce also cooperated in lobbying while organizing pro-nuclear events around the country (Tuuri 2005, 271–5; Hylkilä 2003, 92).

The new lobbying strategy was based on a meticulous information strategy. Two women, Ulla Sirkeinen, a director of the Confederation of Finnish Industry and Employers, and Anneli Nikula from TVO, both graduate engineers, were chosen to act as public advocates (Tuuri 2005, 271–5). The importance of who was delivering the message was emphasized to such an extent that in January 2002, when the government took the DiP, the main pro-nuclear lobbyist from the 1993 nuclear debate, Matti Putkonen of the Finnish Metalworkers' Union, was replaced immediately prior to a television debate by a female advocate, Ulla Sirkeinen. In total, by this time four women were publicly promoting the construction of the new unit, since in Spring 2000 the pro-nuclear MP Sinikka Mönkäre (Social Democratic Party) became the Minister of Trade and Industry, replacing the nuclear-critical MP Erkki Tuomioja from the same party. The fourth was the chair of the Economic Committee Leena Luhtanen (Social Democratic Party) (Kojo 2004, 242, 253). The clear intention of the pro-nuclear lobby was to soften the images of lobbying, especially in the eyes of Finnish women, who were known to be more critical than men of the further construction of NPPs.

In both public and Parliament the pro-nuclear lobby was performed only by the small group of specially selected speakers. This seems to be one of the main contrasts with the pro-nuclear lobby in the early 1990s, when the lobbying lacked such coordination (Ruostetsaari 1998, 154). The strategy also stated that some of the most enthusiastic pro-nuclear individuals were not allowed to express their views in public because the intention was to avoid exaggerated statements. Even the statements of the top industrialists were tightly monitored (*Prima* 2007). According to Matomäki's biography the MP Timo Kalli (Centre Party) was the key figure in Parliament, acting as a focus for the group of pro-nuclear MPs from different political parties. The industrial lobby groups kept up a regular supply of information and arguments that could be used by MPs. One example of the changed strategy of the pro-nuclear lobby was in its attitude to wind power. In 1993 the potential of wind power had been mocked, but on this occasion the lobby adopted a positive attitude. Pohjolan Voima Ltd, the principal owner of TVO, carried out studies on wind power, although the company had little belief in its potential as a form of energy production. All of these were intended to achieve a so-called 'green-wash' of the company (Tuuri 2005, 273–5; Hassi 2002, 315). Instead of mockery, on this occasion the strategy of advocating the promotion of all forms of energy production. Thus, the acquisition of the fifth NPP unit was framed as being environmentally friendly or, to use the label of Lammi (2004, 12–13), as 'the environmental plan of heavy industry'.

Hylkilä (2003, 78), who examined the MPs' perceptions of lobbying around the fifth NPP, concluded that on this occasion the starting points of the pro-nuclear lobby were the dissemination of information, objectivity and openness. However, the core group of four pro-nuclear MPs decided not to act as a group in public. Lammi states that pro-nuclear campaigners were more experienced, had a wider transmission of accumulated knowledge and managed to create a better network both inside Parliament and throughout Finnish society. One example of the wide societal networking of lobbying is the Päivänsäde (Sunshine) project coordinated by the Economic Information Office.[8] In the second half of the 1990s the project focused on the dissemination of information about radiation to Finnish teachers and schoolchildren (Laurila 1995). Such campaigns may have had some impact on the general attitude to nuclear power and nuclear waste management, but, unfortunately, no comprehensive evaluations are available. However, the fact that the project focused on schools indicates just how important the schools were considered to be in the overall societal debate. The nuclear industry sees schools as an arena to reach, offering them a chance to cooperate with and influence the decision makers of the near future (see Ämmälä and Ryhänen 1996, 201). This indicates just how far-reaching the objectives of information programmes and lobbying can be.

Why were those NGOs who opposed the new NPP unit unsuccessful in the lobbying of decision making in 2000–02? In the analysis of this case Lammi considers both internal and external explanations. The anti-nuclear movement had lost many of its supporting network actors – including, for example, many activists in political youth organizations and local trade unions. For example, in the youth organization of Social Democrats many passionate debates on nuclear power with demands for it to be passed out were presented in the late 1980s (see Laakso and Åberg 2006, 319, 323, 344–6).

Furthermore, between 1995 and 1999 there was an almost complete discontinuity of the organizational transmission of campaign narratives in the NGO nuclear campaigns. After Parliament rejected the DiP in September 1993, from the organizational perspective the movement faded away in under two years. The Association for Energy Policy – Alternative to Nuclear Power (Energiapoliittinen yhdistys – vaihtoehto ydinvoimalle [EVY]), one of the most central actors of the Finnish anti-nuclear movement since the late 1970s, decided to suspend its activities in Spring 1995. A new group, 'No More Nuclear Power' (Edelleen ei ydinvoimaa), was organized in Spring 1997, in response to the increased demands of pro-nuclear actors that a new unit should be constructed (Nissinen 2004, 88–9). Thus, the anti-nuclear movement reacted at an early stage to the signals of the reborn pro-nuclear lobby, but the movement only gained significant strength after TVO had applied for a DiP in November 2000. Furthermore, the older and younger generations of the movement favoured different styles of lobbying.

The anti-nuclear movement had close connections to the parliamentary group of Greens, who had planned their own strategy for rejecting the further construction of nuclear power. Their strategy was based on arguments that climate policy and future energy needs could be satisfied through a combination of increased energy efficiency and the development of renewable energy sources (Tontti 2007, 218–19).

Lammi states that a profound change of campaign style took place compared to 1993. There was a strategic change in the arguments presented against nuclear power, with communication being moved away from the risks of nuclear power to the benefits of renewable energy. Campaign messages were focused on the technical potential and economics of renewable energy sources and the arguments were based on MTI's and VTT's reports to ensure credibility.

The re-styling can be seen as a reaction to the traditional image of environmental activists. According to Lammi, the background of the changing style can be traced back to Dodo, the environmental NGO which supplied most of the core group of anti-nuclear activists. Re-styling aimed to reinforce the movement's political credibility, because there were fears that the anti-nuclear movement might otherwise become marginalized. Jouni Nissinen (2004, 91), one of the architects of the renewed movement, concludes that there had been a change in the nature of the arguments used against nuclear power. The earlier fear of radiation was perhaps one of the main arguments, but the younger generation no longer focused on this issue. The re-styling of anti- and pro-nuclear lobbying indicates that both parties had changed their knowledge base and risk perception. These issues are discussed in the final section.

Lammi questions the function of ecological democracy and the participation of civil society in environmental policy making in Finland. He raises the question of who will provide the information to support the other side of the environmental debate, arguing that in the Finnish nuclear debate this task was left to the NGOs. During the debate about the new NPP unit the MTI funded 13 NGOs – both pro- and anti- – with overall funding of €43,000 for information dissemination in 2001 (Kojo 2004, 249). According to Lammi, the funding from the MTI indicates that there was a recognition of the imbalance of resources between the anti- and pro-nuclear lobbies and the small grants to Finnish civil society served mainly as a means of fuelling the societal nuclear debate that was seen as an important aspect in terms of legitimizing the decision-making process. Nevertheless, it is clear that NGOs were poorly funded in comparison with the resources of the pro-nuclear lobby. The anti-nuclear movement was confronted by a series of utopian visions related to the development of nuclear power: trust in technology and its ability to produce welfare; a Finnish political culture consisting of a legalist tradition; corporatism; the strong position of administrative bodies; and the involvement of academic intellectuals in the creation of the national ideology (Miettinen and Väliverronen 1999, 13). Lammi concludes that both the production of decision-making material and the institutional solutions for providing balanced information should be subjects of further study. Lammi's conclusions go to the very heart of nuclear policy decision making, namely the circles of the Finnish nuclear policy establishment: the activity of pro-nuclear politicians (such as the Group of Four MPs), the MTI responsible for the preparation of decision making, the role of state research institutions (such as VTT) and their commercial connections with the utilities and, finally, the role of Finnish mainstream media as watchdogs of societal decision making.

Site selection for the repository for SNF

Site selection for the repository for SNF has been the one of the major unsolved problems of the nuclear industry. In Chapter 6 of this volume the author describes the siting process that is undertaken in Finland. Finnish nuclear waste policy was based partly on reprocessing until September 1995, when the MTI issued the decision regarding the nuclear waste management of IVO. Only since this time have the nuclear power companies TVO and IVO, implemented, following their establishment of Posiva, a consistent nuclear waste policy characterized by national and moral responsibility to take care of SNF by disposing of it in Finnish bedrock. The final shipment to Russia of SNF from the Loviisa NPP took place in December 1996.

The author concludes that the site selection strategy in Finland has changed since the earliest days of the siting process in the early 1980s. The strategy applied was systematic, for geology played an important role in the selection of sites. Although the opinions of the municipalities surveyed were elicited on the conduct of bedrock investigations, in 1992 the selection criteria were still based on geology. Only at the end of the 1990s was the local population of the candidate sites involved in the siting process as part of the implementation of the environmental impact assessment (EIA) process. However, discussion was focused on environmental impacts instead of site selection, which was the main concern of local residents. The new, more interactive style of the waste management company aimed to achieve local acceptance (Kojo 2005; Hokkanen 2007; see also Ämmälä and Ryhänen 1996; Vira 2006).

Since the DiP of 2001 there has been a change in the situation, as the newly established company Fennovoima Ltd submitted an application for a new NPP unit in January 2009 and as part of the application the company presented its plan for SNF management.[9] Fennovoima is naturally willing to negotiate with Posiva on joint measures in respect of the disposal of SNF, but the owners of Posiva, the operating utilities TVO and Fortum, which have also had their own plans for further construction have to date even been unwilling to negotiate on co-operation with Fennovoima. Thus, the national nuclear waste policy seems to be rather fragile, as the existing utilities are able to affect the course of a Finnish nuclear waste policy that is to have one joint repository for SNF produced in Finland. A joint repository was also the main reason for the establishment of Posiva. Furthermore, in its DiP of 2000 the Council of State concluded that 'it is acceptable as well as technically and economically reasonable to concentrate the disposal of spent nuclear fuel produced in Finland in one locality' (Council of State, 2000, 12).

The unwillingness of Posiva to share the cave in Olkiluoto has led to two different interpretations of the Finnish Nuclear Energy Act (Nuclear Energy Act 990/1987). Fennovoima is aiming to establish a system of joint nuclear waste management with Posiva, in accordance with Section 29 of the Nuclear Energy Act, stating, under the waste management obligation, that the ministry may order various licensees to undertake waste management measures on a joint basis (*Taloussanomat* 18 June, 2 November 2007). Posiva's interpretation, however, is different, as the company interpreted that Fennovoima is to become a licensee under the waste management obligation only when Fennovoima has a nuclear reactor with an operating licence (*Kauppalehti* 28 September 2007). The managing director of Posiva, Eero Patrakka, stated that Onkalo (the cave) will not be the disposal facility for the whole of Finland. According to Patrakka, each nuclear power company must be responsible for organizing the disposal of the SNF of its own NPP (*Länsi-Suomi* 2 November 2007).[10] In practice, this would mean that Fennovoima would need to secure a positive DiP and construction and operating licences before the negotiations on the joint nuclear waste management with Posiva could take place. And according to the Nuclear Energy Act, in order to obtain the DiP Fennovoima needs to present its plan for SNF management. Posiva's interpretation on becoming a licensee under the waste management would thus mean that Fennovoima needs to base SNF management around the planning of a new repository in Finland. With the capacity of two repositories Finland would be an even more interesting host for the disposal of SNF.

The strict interpretation of Posiva on joint national policy raises the question of whether or not Posiva is trying to achieve a monopoly over the disposal of SNF in Finland. Such a monopoly would safeguard the interests of both TVO and Fortum in the further construction of nuclear power, as they would hold the disposal card that is vital in securing the licensing of a new NPP. In January 2008 Posiva announced that the company would begin an EIA process for a more extensive repository to fulfill the future disposal needs of its owners. The DiPs of 2000 and 2002 cover 6,500 tU. The approval of Posiva's DiP applications of 2008 and 2009 for the facility extension would increase the maximum dispose capacity to 12,000 tU.

The case of Fennovoima shows how quickly the arguments for final disposal can change, as it was only at the turn of the millennium that Posiva was emphasizing the importance of making a national decision about the disposal of SNF produced in Finland (Posiva 1999, Appendix 2). It seems that Posiva and its owners are not going to show any solidarity towards Fennovoima trying to enter the Finnish nuclear industry. However, the new company has considerable know-how in respect of disposal into bedrock through one of its owners, E.ON, which owns a part of the Swedish nuclear waste company Svensk Kärnbränslehantering AB. Thus, a whole new era would begin if Fennovoima decided to apply for a DiP for a new disposal facility for SNF in Finland. However, Fennovoima will initially try to secure its share of Posiva's facility because that would be the most cost-effective way of organizing the management of SNF. In contrast to the earlier days of nuclear waste management, on this occasion the utilities are struggling against each other – not against the local groups – over the issues of the disposal site and capacity.

To date, the Finnish site selection process has been described as a success story, with acceptance being gained at both the local and national levels. Chapter 6 illustrates how economic aspects – such as tax revenue and compensation – played a crucial role in local decision making where the management of TVO and some active local politicians had gradually enhanced the partnership between the company and the municipality of Eurajoki since the early 1990s. How then can the national decision making be understood? In Chapter 3 Lammi states that the Finnish Parliament decided on 'a testing permit' in 2001 when the DiP of the repository was ratified. The main point was that there was no opposition to further investigations, and that thus Posiva was allowed to start the excavations in Olkiluoto. Even the Greens approved of the application, following the reduction in disposal capacity. The Greens, who were one of the parties of government, were not ready to oppose the final disposal, although the Minister of the Environment, Satu Hassi, tried to delay the DiP's schedule. Raittila and Suominen (2002, 112–13) concluded that the suspicious MPs calmed down both themselves and the voters by converting the DiP to a permission for investigation. The 'least bad' SNF management option was accepted. The Finnish media settled the questions that were posed in the parliamentary debate, but which remained unanswered by writing in terms of 'an interim order' and 'a permission for investigation' (Raittila and Suominen 2002, 113).

Lammi indicates just how the interpretation of final disposal changed over the course of a single year. During the nuclear debate of 2002 the Minister of Trade and Industry, Sinikka Mönkäre, presented the ratification of the DiP as a solution to the problem of nuclear waste, whereas in 2001 it had focused on the testing permit. The problem of nuclear waste literally faded away, for following the vote on the NPP unit taken in May 2002 Parliament agreed without any further discussion not to speak of voting on the extension of the facility for the SNF produced in the new NPP unit.

Knowledge production, selection and risk perception

Chapter 4 on MPs' thoughts regarding the decision on the fifth NPP unit sheds more light on the discursive middle ground based on the ideas of ecological modernization. This new middle ground has reconciled the old contradiction between simple and reflexive modernization. Berg states that while reacting to environmental risks was widely supported by MPs, criticisms of modern growth logic have not been adopted in wider circles. Furthermore, the construction of NPPs was no longer defended simply in terms of 'hard' economic and opposed in terms of 'soft' environmental argumentations, as shown in Chapter 3. The bipolar framing of the nuclear debate was reformulated particularly as a result of changes in the politics of climate change. In the eyes of some MPs the promotion of nuclear energy was thought to safeguard the economic growth needed for the maintenance of the Finnish welfare state. Furthermore, this growth was seen to be based on cleaner electricity production because nuclear energy was seen as being free of CO_2 emissions. Berg has labelled this phenomenon the discourse of combating climate change with nuclear energy.

Berg concludes that the decisive battle over votes among Finnish MPs took place in the middle ground where the logic of ecological modernization was utilized both to defend and to oppose the construction of the new NPP unit. MPs' views on environmental risk were reframed as the impact of the Chernobyl accident, albeit severe, was shown to be both short term and localized compared to that of climate change. The case of Chernobyl became normalized and the temporary distrust in nuclear technology and expert knowledge that it had engendered changed gradually into a more optimistic outlook. The former Prime Minister of Finland, Paavo Lipponen,[11] offers a clear example of just such a changed perception. Following the Chernobyl accident, Lipponen (1986, 105–6) outlined an energy policy in which there was to be no construction of any new nuclear capacity. According to Lipponen, the phasing out of nuclear energy within a reasonable timeframe would not lead to an economic or environmental catastrophe. However, he predicted that industry and the power companies would not readily abandon the further construction of nuclear power plants. And if nuclear energy was to be reconsidered then, according to Lipponen, this should be done on the basis of a broad consensus based on a referendum. During his time as Prime Minister Lipponen defended the construction of an additional NPP unit. Following the parliamentary vote of 2002 Lipponen stated that the calm decision making was in part due to the comprehensive accounts made of the Chernobyl accident (*Satakunnan Kansa* 25 February 2002).

The reframing of nuclear risk was influenced at least in part by the fact that since the debate in the early 1990s the strategy of the anti-nuclear movement had also become less focused on the idea of the nuclear hazard. In addition, there was a strong degree of trust in Finnish technology. The control of risks was seen to be in the hands of experts who, at least in the case of Finland, had carried out their duties in a responsible manner. Nuclear power as a safe energy option for both society and the environment was justified on the grounds of Finnish know-how and their well-defined sense of social conscience (Halme and Takala 2003, 513). Thus, in the Finnish debate nuclear technology had a strong nationalistic feature, which the applicant TVO attempts to enforce by sponsoring, for example, one of the most popular sports in Finland – ice hockey. National identity was used by the pro-nuclear lobby in Finland, especially in their warnings against a dependence on the import of electricity from Russia – from the old Russian NPPs. Thus, in this debate the further construction of *Finnish* nuclear energy was contrasted with domination by Moscow. Interestingly, in late communist societies in the late 1980s the anti-nuclear movement had used similar objections to domination by the old Soviet overlords as an argument *against* the expansion of nuclear power in their countries (Dawson 1996). In the post-communist societies, however, public attitudes seemed to become increasingly pragmatic. Opposition to nuclear power was viewed as undermining attempts of the newly independent state to survive economically (Dawson 1996, 176). The same types of arguments can be identified in the Finnish nuclear debate. Lammi (2004, 46–7; see also Halme and Takala 2003, 505–6; Jarva 2006) concludes that in the nuclear debate of 2000–02 the Finnish MPs clashed over societal storylines. Thus, the pro-nuclear attitude was associated with the defence of the welfare state and employment, whereas the anti-nuclear attitude was associated with the spin of unsustainable western industrial policy. Stories were used by players both to explain and to legitimate their actions. According to Lammi (2004), stories were a struggle over societal power triggered by the parliamentary rejection of 1993.

Berg gives a clear description of how the members of the Finnish Parliament were the focus of knowledge production. In the case of the fifth NPP unit the volume of information confused at least some of the MPs, although at the same time the very fact of being a target for knowledge production seemed to have enhanced their status as decision makers of a representative system. This identity served to confirm their view of badly informed citizens who were incapable of handling the mass of information and were also unable to make well-informed independent decisions, thus making a referendum unnecessary. Although we can give no detailed analysis of supporting factors for the decision making of individual MPs, Berg concludes that in general MPs found at least the expert hearings in the committees, their peer groups and their own political views to be supporting sources in decision making. Hylkilä (2003, 90–3) describes how the pro-nuclear lobby was capable of using its peer groups (both inside and outside Parliament) as one important means of influencing individual MPs. No doubt the lack of continuity had damaged the network of personal relations within the anti-nuclear movement, thus making it much more difficult to achieve large-scale peer group lobbying. Furthermore, the movement was poorly resourced.

One interesting finding, which offers a prime subject for further research, was the process in which a view presented, for example in the parliamentary hearings, was interpreted as biased, thus undermining the 'expert' status. From the point of view of lobbying, it is vital to understand how the idea of 'expert' status is affected. Berg's interviews affirmed Hylkilä's (2003, 136–8) finding that a greater degree of attention was paid to the views of those experts with a strong institutional background, such as university professors and the representatives of individual ministries. It remains unclear to what extent the MPs were ready and able to reflect the knowledge production chain remains unclear.

One interesting example of knowledge selection can be found in the preparatory process of the DiP concerning the final disposal facility of SNF when STUK made reference to the External Review Group Consensus Report. The summary of the report modified by STUK was less critical than the wording of the original report. The members of the Parliament were given access to the Finnish summary, but not to the original text (Lammi 2004, 37). Lampinen (Chapter 2) states that the MTI can guarantee the smooth handling of the applications in the face of criticism. Thus, the most important authority in respect of Finnish nuclear energy policy, the Ministry of Trade and Industry, is criticized for its strategic – or even manipulative – action. The same kind of criticism was voiced as a result of research in which the written opinions of Posiva's EIA process were analysed (Hokkanen and Ruuskanen 2005, 75–6; Hokkanen 2007, 183–202). Once again, power in Finnish nuclear energy policy seems to have been concentrated in the hands of the MTI. Although the criticism is based on a few examples, it reveals how power is exercised in Finnish nuclear policy. The conclusion drawn from the criticism is that there are still serious problems with stakeholder involvement in the Finnish decision-making process. Both Lampinen and Lammi present examples that highlight the lack of transparency and the poor quality of the existing review process.

A shift of legitimate knowledge occurred among the core activists of the anti-nuclear movement. Lammi details how the arguments and the political style of the anti-nuclear movement changed principally because a new generation of activists took charge of the movement. Lammi's analysis indicates that the formation process of a piece of knowledge that is relevant to the development of energy policy is dependent upon many coincidental factors. In a political debate which has at least two conflicting viewpoints, the life-cycle of politically relevant knowledge can change dramatically when one of the main actors decides to reform its own strategic action, including its fundamental arguments.

For example, the debate around the new unit under construction in Olkiluoto indicates that nuclear safety arguments have now returned to the agenda. At the beginning of April 2007 Greenpeace demonstrated on the coast in front of Olkiluoto Island. The arguments advanced by them concerned the reported safety failures of the EPR unit listed by STUK.

According to Kantola (2007), a piece of legitimative knowledge emerges once a set of pieces of fixed technical knowledge have been successfully translated and combined into a cogent whole. A large part of the diffractions of the discourse have been 'black-boxed' off from the public. In part, this is a characteristic of any research process, but the point is to note that there are several mutually exclusive ways of combining the relevant pieces of technical information into a map of orientation for the future. The value-laden combinations and, in particular, the political decisions of choosing one of them to be the road map of society should be made transparent. One interesting point is how the commitment of the investors in research programmes is able to regulate the outcome of the negotiations. None of the actors in the negotiation network can decide solely and independently on the content of the knowledge. However, an interesting question is how much a consistent establishment can self-regulate the outcomes of the negotiations on legitimative knowledge.

Based on the practical observations and the theoretical input of the chapters it can be concluded that in a strong administrative state, such as Finland, the political steering ideology is hidden in the regulatory measures of the state. Thus, neither political power nor political responsibility is in the hands of the political parties (as has been the case with strong party states), but the power is taken by the faceless players in the political establishment.

In Chapter 7 Litmanen examines Mary Douglas's cultural risk theory and its deficiencies in order to develop adequate explanatory tools to conduct an analysis of the Finnish case. According to Douglas, we choose risks simultaneously with our choice of social institutions. However, Litmanen argues that this kind of analysis overlooks the explicit role played by powerful societal actors and institutions, such as government and the media. According to Litmanen, these actors may even have some causal power over either people's risk perception or the predominant societal risk assessment. Thus, Litmanen seeks to develop Douglasian cultural risk theory in order that it also covers both contextual factors and societal processes. In order to understand the complex dynamics of societal risk evaluation, the contextual and embedded nature of individual and collective interpretations need to be connected to the relevant societal, historical and geographical contexts.

Litmanen explains the Finnish nuclear decisions with the assistance of contextual factors and broader societal developments that have taken place over the course of the past decade or so. According to Litmanen, the national peculiarities and features of the Finnish decision-making system are preconditions of societal decision making, but not to a sufficient degree to explain the content of the decisions. Litmanen concludes that Finland has undergone a transformation since the early 1990s when the Parliament of Finland rejected the DiP application for a new NPP unit. Litmanen points to the renewal of the country through an analysis of changes in media, lobbying, science, values (such as environmentalism), societal structures and national strategy. In the early 1990s the decision to reject the NPP application was characterized, according to Litmanen, by strong economic concerns and social insecurities. The emphasis on ecological sustainability and the impact of the spirit of the Rio Summit of 1992 on Finnish energy policy could also been seen as forming part of the background for the rejection (Kyllönen 2004, 54–5). However, uncertainty and instability also played a role during the early years of the twenty-first century. Litmanen suggests that new threats, such as 9/11 and the future direction of Russian society, have also had an effect on nuclear decisions.

Explaining the revival of nuclear energy in Finland

The main purpose of this chapter has been to find out why Finland has experienced a revival of nuclear energy in recent years. This revival can be seen to have culminated in the ratification of the DiP by 107 votes to 92 in the Finnish Parliament in May 2002. Paavonen (2006, 91) concluded that as a result of the decision Finland had returned to an energy policy that supported economic growth. According to Paavonen (2006, 92), in 1993, when the DiP was rejected in Parliament by 107 votes to 90, the decisive factor was that the majority of the Social Democratic Party group voted against the further construction, whereas in 2002 the majority of them voted in favour in accordance with the opinion of the chair of Social Democratic Party, Prime Minister Paavo Lipponen. In 2004 the author (Kojo 2004, 239) concluded, firstly, that by 2002 Finnish public opinion were less worried about the threat of a serious nuclear accident such as the Chernobyl disaster. Secondly, the further construction was framed by the pro-nuclear lobby in terms of an environmental project helping to reduce the level of CO_2 emissions. Thirdly, the ratification of the DiP concerning the disposal of SNF in 2001 was also interpreted as a solution to the problem of nuclear waste. Fourthly, the strategies of pro- and anti-nuclear coalitions changed. Furthermore, some larger societal changes, such as the scientization of lobbying, were identified as a background to the favourable decision. What other aspects have been presented to explain the decisions? The previous literature on the situation in Finland has offered few insights. In what remains the main findings of the chapters are discussed, along with some conclusions from the existing research.

Hakapää (2004) and Yamasaki (2007) have both applied a policy approach to their study of Finnish nuclear energy decisions. Hakapää (2004, 23) states that the Finnish decision provides us with 'a chance to examine the ties but also contradictions between European, regional and national energy policies in a situation when a state gives a considerable support for European and regional co-operation but naturally keeps its own national aims in mind too'. His conclusion is that the nuclear option was justified by the state and the industry on the grounds of competitive prices, the security of energy availability and the fulfillment of international environmental agreements (Hakapää 2004, 26–30). Finland chose not to take the further step of relying on European or regional energy policy projects. Thus, according to Hakapää (2004, 34), sovereignty became the most important principle underlying the formulation of Finnish energy policy. The rise of nationalistic features was also noted by Litmanen, who states that the parliamentary decision of 2002 to approve the further construction of nuclear power was influenced by the end of the era of what Litmanen has called 'benevolent globalization'. The new millennium offers new challenges, for example, with the developments in Russia and the possible oil crisis that could result from the War on Terror, which has swept away much of the optimism linked to globalization after the end of the Cold War. Naïve neo-liberalism was transformed into realism, helping the pro-nuclear lobby to sell the new NPP as a project supporting national identity (see Chapter 7; Litmanen 2004, 228–9).

Yamasaki (2007, 18) concluded on the basis of the Finnish research that nuclear industry lobbying appeared to explain much of the positive reaction by the government to the demand for the life extension of NPPs and to the 2002 granting of the DiP for the fifth NPP. The governmental decision of 1998 to grant the extension of Loviisa NPP is presented by Yamazaki as a major continuation of policy. According to her, in terms of the development of Finnish nuclear energy policy, it is the 1993 decision that can be seen as the exception. Hakapää (2004, 30) also describes how the strong dislike of nuclear power was changed as the result of the rational and pragmatic campaign for the positive image of nuclear power. The differences from the deleterious and negative campaign of 1993 were stark. However, one opinion poll indicated that in Autumn 2002 almost half (46 per cent) of all Finns still mostly or fully disagreed with the statement that a fifth NPP unit should be built in Finland. By contrast, the figure of those fully or mostly agreeing stood at 36 per cent (Kiljunen 2007). By 2003 a shift in public opinion had taken place. According to Kiljunen (2007), before 2003 not one survey had shown a distribution indicating greater support for acceptance than for rejection. The results of 2003–06 deviate visibly from the general trend in the time series that began in 1984. Kiljunen (2004) concluded that the shift was probably a result of relatively straightforward social and psychological adaptation processes – what might be labeled as 'swallowing the facts'. In the area of socially contentious issues, political decision making – its result – influences public opinion. According to Kiljunen (2004), the concept of cognitive dissonance was also appropriate in this interpretation. In order to avoid a conflict between a person's mental and conscious elements, it is easier to adjust one's own attitudes than it is to change the circumstances (Kiljunen 2004). From the point of view of lobbying, this book sheds greater light on the changed strategies of the pro- and anti-nuclear coalitions over this period. However, there is still a need for more specific studies on the informal connections between the main actors in order to achieve a better understanding of the real nature of the Finnish decision making behind the formal processes.[12]

Based on the conclusions of the authors and the policy arrangement approach (Arts and van Tatenhove 2004), the following summary can be drawn. First of all, the pro-nuclear policy coalition, TVO, its industrial owners and some politicians at the core of the coalition, were more powerful than the Finnish anti-nuclear movement. The pro-nuclear establishment had lobbied for the further construction of nuclear power for the whole decade after the defeat of 1993. The pro-nuclear establishment was more experienced and had broader societal networks for lobbying than did the anti-nuclear movement, which was suffering from discontinuity and a lack of both monetary and non-monetary resources. In addition, the public–private partnership between the applicant companies and the authorities provides an example of a strong policy coalition. Secondly, although the anti-nuclear movement could influence the nuclear policy discourses, it lacked the political muscle to reformulate energy policy. Thus, the rules of the game, which are in part implicitly described by the report of the World Energy Council (2007, 72–5), were set principally by the pro-nuclear policy coalition. For example, the anti-nuclear movement had serious difficulties in influencing preparation processes because the MTI ignored repeated calls to produce a third scenario for climate policy.

Moreover, the nature of the legal-administrative framework for the decision-making procedure cannot be ignored in any discussion of the Finnish rules of the game. Chapter 5 by Säynässalo offers the first comprehensive comparative study of the Finnish decision-making procedure in respect of nuclear energy policy. She raises the question of the lack of capacity of Finnish party politics to steer nuclear energy policy. The legal-administrative framework, which has now been tested twice in NPP projects, and once with the SNF facility has resulted in the granting of both favourable and unfavourable decisions. Thus, although the basic nature of the DiP procedure remained the same between 1993 and 2002, the differences in terms of context, strategies, players, discourses and arguments led to different end results. These changes have been analysed in the chapters of this book.

Can Finland then be considered a suitable example for other countries to follow in respect of the development of new nuclear power programmes? Finland is used – and will continue to be used – as an example of combining nuclear energy and a democratic decision-making process (see, for example, Riley 2004, 83; Grimston 2005, 43–6; Lipponen 2007). Although there may be attempts to 'play the Finland card', one should bear in mind the particular Finnish sociopolitical realities presented in the chapters (see also Grimston 2005, Appendix 1) and thus the difficulty and uncertainty of adapting such context-dependent practices to different societal circumstances.

Notes

1. At the time of finishing this chapter, TVO, Fennovoima and Fortum have all applied for a DiP for a new NPP unit.
2. The MTI was integrated into the Ministry of Employment and the Economy. The new ministry was established on 1 January 2008.
3. The Advisory Committee is appointed by the ministry.
4. The motives of the Finnish Greens are studied by Darst and Dawson (2007) in a conference paper comparing the intersection of the issue of SNF disposal with that of nuclear power generation in Germany and Finland.
5. At the time of submitting Posiva's application in May 1999 both of Posiva's owners, TVO and Fortum, were implementing EIAs of their own for a new NPP unit. Thus, Posiva included in the application the amount of SNF produced by four operating units and two units which were at the planning stage. In Autumn 1999 TVO and Fortum decided that TVO would be in charge of the new NPP unit process.
6. Criticism was expressed as part of the proposal of Finland's energy strategy in 1991. In their dissenting opinion three members of the parliamentary Energy Policy Committee criticized the fact that the role of Parliament was minimized in the Nuclear Energy Act. They were critical that Parliament was missing possibilities to set preconditions, to reformulate the content of the Act and to postpone the decision over the next elections. Parliament is normally vested with these rights in the preparation procedure of an Act (Energy Policy Committee 1991, 27).
7. Between 1999 and 2001 Tauno Matomäki was the chairman of the forest industry company UPM-Kymmene's Board of Directors. UPM-Kymmene is the largest shareowner of PVO which in turn owns more than 50 per cent of TVO.
8. The Economic Information Office is a part of the TAT Group owned by the Confederation of Finnish Industry and the Employers' TT Trust and Confederation of Finnish Industries.
9. Fennovoima announced its NPP construction plan in June 2007. According to the company, 30 sites were given a preliminarily evaluation by the beginning of Summer 2007. By the end of 2007 Fennovoima had announced that the company would initiate an EIA process for an NPP in the municipalities of Kristiinankaupunki, Pyhäjoki, Ruotsinpyhtää and Simo. In June 2008 Fennovoima excluded the site of Kristiinankaupunki. None of the municipalities under consideration have any previous experience on nuclear facilities. However,

Ruotsinphytää is located next to the nuclear community of Loviisa. Both municipalities will be consolidated into the new town of Loviisa in 2010.

10. The senior manager of Posiva, Timo Seppälä, went even further as he stated that '[P]erhaps it's reasonable that Fennovoima constructs a repository of its own. It seems like that more nuclear power will be constructed. Surely it would be sensible, that another repository was located in Finland' (*Seura* 7 March 2008).
11. Since August 2007 Paavo Lipponen has worked as a political advisor to PVO in climate and energy issues.
12. For this kind of study the biographies and memoirs of politicians and other major players are a valuable source. The memoirs of the former Minister of the Environment Satu Hassi (2002) is an excellent example. By contrast the memoirs by the former Prime Minister Esko Aho (1991–95) and the Minister of Finance Sauli Niinistö (1996–2003) were disappointing from the point of view of studying the Finnish nuclear debate and decision making. In the near future the memoirs of the former Prime Minister Paavo Lipponen (1995–2003) will hopefully be more enlightening.

References

Advisory Committee on Nuclear Energy (1998) *Kansallinen ydinvoimalaitosten turvallisuustutkimus 1999–2002: ehdotus uuden tutkimusohjelman sisällöksi ja organisoinniksi* (National Nuclear Power Plant Safety Research 1999–2002. A Proposal for the Content and Organisation of the New Research Programme). Studies and Reports 15/1998. Helsinki: Ministry of Trade and Industry.

Ämmälä, V.-M. and V. Ryhänen (1996) Experiences from TVO's Communication Programme on the Siting of Spent Fuel. In *Informing the Public about Radioactive Waste Management. Proceedings of an NEA International Seminar, Rauma, Finland, 13–15 June 1995*. Nuclear Energy Agency, Paris: OECD, pp. 197–202.

Arts, B. and J. van Tatenhove (2004) Policy and Power: A Conceptual Framework Between the 'Old' and 'New' Policy Idioms. *Policy Sciences* 37(3–4), 339–56.

Council of State (2000) *Valtioneuvoston periaatepäätös*. (The Decision in Principle concerning the final disposal of spent nuclear fuel produced in Finland). M 7/2000 vp. www.eduskunta.fi.

Darst, R. and I. Dawson (2007) *Beyond Constructive Procrastination? Radioactive Waste Disposal and Nuclear Power in Germany and Finland*. Paper prepared for the 48th Annual Convention of the International Studies Association, Chicago, 28 February–2 March.

Dawson, J.I. (1996) *Eco-Nationalism: Anti-nuclear Activism and National Identity in Russia, Lithuania, and Ukraine*. Durham and London: Duke University Press.

Energy Policy Advisory Council (1972) *Suomen energiantarve 1970-luvulla: Energiapolitiikan neuvottelukunnan mietintö* (Finland's Need of Energy in the 1970s). The Committee Report A, 1972, 11. Helsinki: Ministry of Trade and Industry.

Energy Policy Committee (1991) *Suomen energiastrategia: Energiapolitiikan neuvoston ehdotus* (Finland's Energy Strategy: The Proposal of the Energy Policy Council). The Committee Report 1991: 29. Helsinki: Ministry of Trade and Industry.

Elo, M. (2007) *Politiikkaa ja persoonia* (Politics and Persons). Helsinki: Etukeno.

Frogatt, A. (2007a) The Status of Nuclear Power in Europe and the Impact of Economics on its Future Development. Presentation in *the European Nuclear Critical Conference*, Helsinki, 10 November. http://www.uraanitieto.tormunet.fi/encc/index_eng.htm.

Frogatt, Antony (2007b) Nuclear Power – The European Dimension. In D. Elliott (ed.), *Nuclear or Not? Does Nuclear Power Have a Place in Sustainable Energy Future?* New York: Palgrave Macmillan, pp. 171–82.

Grimston, Malcom (2005) *The Importance of Politics to Nuclear New Build: An Examination of the Relationship between Political, Scientific and Public Mindsets and its Influence on Decision-making in the Scientific and Technical Field.* London: The Royal Institute of International Affairs, Chatham House. http://www.chathamhouse.org.uk/files/3290_dec05nuclear.pdf (accessed 21 September 2007).

Hakapää, J. (2004) Finland's Choice for Nuclear Power in Context of National, Regional, and European Energy Policies. In M.-B. Schartau (ed.), *Energy Resources, Energy Policy and Democratic Development in the Baltic Sea Region.* Gdansk-Berlin: Wydawnictwo Uniwersytetu Gdanskiego, Nordeuropa-Institut der Humboldt-Universität zu Berlin, pp. 19–43.

Halme, P. and T. Takala (2003) Vastuullinen suhtautuminen ydinvoimapäätökseen – diskurssianalyyttinen tutkimus (Responsible Attitude to the Decision on Nuclear Power – a Discursive Analytic Study). *Liiketaloudellinen aikakauskirja.* 4, 495–517. http://lta.hse.fi/2003/4/lta_2003_04_a3.pdf.

Hassi, S. (2002) *Tukka hattuhyllyllä* (Hair on the Hat Rack). Helsinki: WSOY.

Hecht, G. (1998) *The Radiance of France: Nuclear Power and National Identity after World War II.* Cambridge, MA: The MIT Press.

Helsingin Sanomat (2006) Interview of Researcher Milka Sunell on 30 April 2006.

Hokkanen, Pekka (2007) *Kansalaisosallistuminen ympäristövaikutusten arviointimenettelyssä* (Public Participation in Environmental Impact Assessment). PhD thesis in Political Science. Acta Universitatis Tamperensis 1285, Tampere: Tampere University Press. http://acta.uta.fi/pdf/978-951-44-7178-0.pdf.

Hokkanen, P. and M. Kojo (2000) *YVA-yleisötilaisuudet kansalaisosallistumisen areenana. Tapaustutkimus TVO:n ja Fortumin lisäydinvoimahankkeista* (Public Meetings of EIA Process as an Arena for Citizen Participation. A Case Study on the Further Construction Plan of Nuclear Power by TVO and Fortum).Working report 1/2000. Department of Political Science and International Relations. University of Tampere.

Hokkanen, P. and T. Ruuskanen (2005) *Kirjallisten mielipiteiden vaikuttavuus YVA-menettelyssä* (The Effectiveness of Written Addresses and Arguments in EIA Report). Alueelliset ympäristöjulkaisut 396. Jyväskylä: Central Finland Environment Center.

Hylkilä, I. (2003) *Suomen viidennen ydinvoimalan lobbaus kansanedustajan näkökulmasta* (Lobbying the Fifth Nuclear Power Plant Unit in Finland from the Point of View of a Member of Parliament). Unpublished master thesis study in Organizational Communication and PR. Department of Communication. Faculty in Humanities. University of Jyväskylä. http://urn.fi/URN:NBN:fi:jyu-2003920889.

Jarva, A. (2006) *Moniulotteisia tulkintoja kestävästä kehityksestä: diskurssianalyysi vuoden 2002 ydinvoimakiistasta* (Multiple Interpretations on Sustainable Development: a Discursive Analysis on the Struggle over Nuclear Power in 2002). Unpublished Masters thesis study in Management and Organisation. Turku School of Economics.

Kantola, I. (2007) *The Role of Knowledge in Energy Politics: A Sociological Study.* Annales Universitatis Turkuensis, Ser. B 298, University of Turku, Turku.

Kauppalehti (2007) Posiva jättää Fennovoiman ulos ydinjäteluolastaan (Posiva excludes Fennovoima from its Nuclear Waste Cave). 28 September.

Kiljunen, P. (2004) *Energy Attitudes 2003: Results of a Follow-up Study Concerning Finnish Attitudes Towards Energy Issues 1983–2003.* Finnish Energy Industries Federation. http://www.sci.fi/~yhdys/eas_03/english/e_eluku2.htm.

Kiljunen, P. (2007) *Energy Attitudes 2006: Results of a Follow-up Study Concerning Finnish Attitudes Towards Energy Issues 1983–2006*. Finnish Energy Industries Federation. http://www.sci.fi/~yhdys/eas_06/english/eng_luku2.htm.

Klötzer, U. (2006) *Säteilevä tulevaisuus* (Radiant Future). Sahlgrens.

Kojo, M. (2005) Changing Approach: Local Participation as a Part of the Site Selection Process of the Final Disposal Facility for High-level Nuclear Waste in Finland. Paper presented in The 10th International Conference on Environmental Remediation and Radioactive Waste Management, Glasgow, Scotland, 4–8 September. Proceedings.

Kojo, M. (2004) Yhteenveto (Summary). In M. Kojo (ed.), *Ydinvoima, valta ja vastarinta* (Nuclear Power, Power and Resistance). Helsinki: Like, pp. 231–58.

Kyllönen, S. (2004) Ydinvoiman ilmastonmuutos (Climate Changes and Nuclear Power). In Kojo, M. (ed.) *Ydinvoima, valta ja vastarinta* (Nuclear Power, Power and Resistance). Helsinki: Like, pp. 51–85.

Laakso, M. and V. Åberg (2006) *Sosialismiin! Sosialidemokraattiset nuoret 1906–2006* (To Socialism! Youth of Social Democrats in 1906–2006). Helsinki: Sosialidemokraattiset nuoret ry.

Lammi, H. (2004) Tarinat kovasta ytimestä (The Stories Concerning the Hard Nucleus). In Kojo, M. (ed.) *Ydinvoima, valta ja vastarinta* (Nuclear Power, Power and Resistance). Helsinki: Like, pp. 11–50.

Laurila, E. (1995) Päivänsäde lisää säteilytietoutta kouluissa (Sunshine Enhances Radiation Knowledge in Schools). In *ATS Ydintekniikka* (24) 3/95, 17–18.

Lipponen, P. (1986) *Muutoksen suunta* (The Direction of the Change). Helsinki: Kirjayhtymä.

Litmanen, T. (2004) Suomen ydinvoimaihme (Finland's Nuclear Power Miracle). In M. Kojo (ed.) *Ydinvoima, valta ja vastarinta* (Nuclear Power, Power and Resistance). Helsinki: Like, pp. 209–30.

Länsi-Suomi (2007) Posiva aikoo huolehtia vain Teollisuuden Voiman ja Fortumin ydinjätteistä (Posiva is Going to Take Care of Nuclear Waste Produced by Teollisuuden Voima and Fortum Only), 2 November.

Malaska, P. and P. Kasanen (1987) *Ydinvoima – kohtalon kysymys?* (Nuclear Power – a Question of Fate?). Energy Department. Series B:63. Helsinki: Ministry of Trade and Industry.

Mansner, M. (2007) *Suurlakosta Euroopan unioniin. Vuosisata työnantajatoimintaa* (From the General Strike to the European Union. One Century of Employer Activity) Helsinki: The Confederation of Finnish Industries.

Michelsen, K.-E. and T. Särkikoski (2005) *Suomalainen ydinvoimalaitos* (Finnish Nuclear Power Plant). Helsinki: Edita.

Miettinen, R. and E. Väliverronen (1999) In Science and Technology We Trust: On the Public Understanding of Science in Finland. In Biotechnology and Public Understanding of Science. Proceedings of the UK–Nordic Co-operative Seminar. Helsinki, 25–7 October 1998. Suomen Akatemian julkaisuja 2/99, 11–22.

MTI (2000) *Toimenpiteitä ydinenergia-alan tietämyksen säilyttämiseksi: työryhmäraportti* (Measures to Retain Expertise in the Nuclear Energy Sector. Working Group Report). Ad hoc committee reports 10/2000. Helsinki: Ministry of Trade and Industry.

NEA (2002) *Stepwise Decision Making in Finland for the Disposal of Spent Nuclear Fuel*. Workshop Proceedings. Turku, Finland. 15–16 November 2001. Nuclear Energy Agency.

Nissinen, J. (2004) Valtavirran vastarinta (The Resistance of Valtavirta). In M. Kojo (ed.), *Ydinvoima, valta ja vastarinta* (Nuclear Power, Power and Resistance). Helsinki: Like, pp. 87–105.

Nuclear Energy Act (*Ydinenergialaki*) *990/1987.*

Paavonen, T. (2006) Teollinen Suomi (Industrial Finland). In T. Paavonen and O. Kangas, *Eduskunta hyvinvointivaltion rakentajana* (Parliament Building the Welfare State). Helsinki: Edita, pp. 88–108.

Prima (2007) Nakkilan Matomäestä suurta lamaa tainnuttamaan (From Matomäki in Nakkila to Knocking out the Great Depression). The interview of Tauno Matomäki. In *Prima* 1/2007, 58–61. http://www.digipaper.fi/prima/3435 (accessed 1 August 2007).

Posiva (1999) *Käytetyn ydinpolttoaineen loppusijoituslaitoksen periaatepäätöshakemus* (Application of Decision-in-Principle to the Final Disposal Facility for Spent Nuclear Fuel). Helsinki: Posiva.

Raittila, P. and P. Suominen (2002) Keskustelu ydinjätteen loppusijoitusta koskevasta periaatepäätöksestä eduskunnassa ja mediassa (Discussion on the Decision-in-Principle on Nuclear Waste Disposal in Parliament and the Media). In P. Raittila, P. Hokkanen, M. Kojo and T. Litmanen (eds), *Ydinjäteihme suomalaisittain* (The Finnish nuclear waste miracle). Tampere: Tampere University Press, pp. 92–113.

Riley, P. (2004) *Nuclear Waste: Law, Policy and Pragmatism.* Aldershot: Ashgate.

Ruostetsaari, I. (1998) *Energiapolitiikka käännekohdassa. Järjestöt ja yritykset vaikuttajina vapautuvilla energiamarkkinoilla* (Energy Policy at a Turning Point. Organizations and Companies as Opinion Leaders on the Liberalizing Energy Market). Research Reports 8/1998. Department of Political Science and International Relations. University of Tampere.

Satakunnan Kansa 25 May 2002. A regional newspaper.

Seura (2008) Lopullinen ratkaisu (The Final Solution). 7 March.

Sunell, M. (2001) *Miten Suomen yksityinen metsäteollisuus hankki länsimaisen ydinvoimalan. Tutkimus taloudellisesta ja poliittisesta vallankäytöstä 1970-luvulla* (How the Finnish Private Forest Industry Acquired a Western Nuclear Power Plant. A study on Use of Economic and Political Power in the 1970s). University of Helsinki. Department of Social Science History. Unpublished Masters thesis study. April 2001. http://ethesis.helsinki.fi/julkaisut/val/yhtei/pg/sunell/mitensuo.pdf.

Tammilehto, O. (1994) Miksi viidettä ydinvoimalaa ei rakennettu? Ydinvoiman vastaisen liikkeen historiaa. (Why Was the Fifth Nuclear Power Plant Unit Not Constructed? The History of the Anti-nuclear Movement). In T. Myllyntaus (ed.), *Liikkeen voimaa – kansalaistoiminta ympäristökysymysten muovaajana* (Power of Movement – Citizen Activity Shaping the Environmental Question). Research report 122. Research Institute of North Finland. University of Oulu.

Taloussanomat (2007) Valtio ei rupea jätekosinnan kätilöksi (The State is Not Going to be a Midwife for Waste Proposal). 18 June.

Taloussanomat (2007) Fennovoima: ydinjätteet kansallinen kysymys (Fennovoima: Nuclear Waste a National Issue). 2 November.

Tontti, J. (2007) Ode-ilmiöstä oppositioon. (From the Ode Phenomenon to Opposition). In T. Remes and J. Sohlstén (eds) *Edellä! Vihreä liitto 20 vuotta.* (In front! The Green League 20 years). Helsinki: The Green Cultural Association, pp. 215–45.

Tuuri, A. (2005) *Tauno Matomäki. Suvipäivänä syntynyt* (Tauno Matomäki. Born on a Summer Day). Helsinki: Otava.

Vira, J. (2006) Winning Citizen Trust. The Siting of a Nuclear Waste Facility in Eurajoki, Finland. *Innovations* Fall 2006, 67–82.

World Energy Council (2007) The Role of Nuclear Power in Europe. January 2007. http://www.cna.ca/english/Studies/WEC_Nuclear_Full_Report.pdf (accessed 29 November 2007).

Yamasaki, S (2007) *Major Changes in Nuclear Energy Policies: A Comparative Analysis of 10 OECD Countries*. Chapter 11, Finland. Draft of PhD Dissertation. 8 August. National Fund for Scientific Research (FNRS Belgium) and Université catholique de Louvain, Center for Comparative Politics.
Yle (2005) Yleisradio's A-Piste television programme, 22 April 2005. http://www.yle.fi/aohjelmat/apiste/arkisto/id15536.html (accessed 13 September 2007).

Appendix 1 The formal decision-making process of the fifth NPP unit in Finland

Time	Milestone
1998	
2 June	EIA programme submitted to the MTI by TVO
29 June–31 August	Public hearing of the TVO's EIA programme
26 October	Statement of the contact authority (MTI)
1999	
30 August	EIS submitted to the contact authority (MTI) by TVO
13 Sept.–11 Nov.	Public hearing of TVO's EIS
2000	
17 February	Statement of the contact authority (MTI)
15 November	DiP application submitted to the MTI by TVO
5 Dec.–28 Feb.	Public hearing of the TVO's DiP application
2001	
7 February	Preliminary safety assessment of the new NPP project by STUK
14 February	Public gathering concerning the DiP application in Eurajoki, arranged by MTI
19 March	Favourable statement of the Municipal Council of Eurajoki
2002	
7 January	Supplement of the preliminary safety assessment by STUK
17 January	Two DiPs by the Council of State concerning the TVO's application for a new NPP and Posiva's application for the extension of the planned disposal facility
13 February	Preliminary debate in Parliament
	Parliamentary expert hearing process in the Committees
24 May	Final reading in Parliament
	DiP ratified by Parliament by 107 votes to 92
2003	
16 October	TVO chose the Olkiluoto site for the new NPP unit
December	TVO chose the reactor type EPR (European Pressurized water Reactor)
2004	
8 January	TVO's application for the building permit
13 Jan.–30 April	Statement process on the building permit
2005	
21 January	Statement of STUK concerning safety
17 February	TVO's building permit approved by the Council of State

Appendix 2 The formal decision-making process of the final disposal facility for the spent nuclear fuel in Finland

Time	Milestone
1998	
6 February	EIA programme submitted to the MTI by Posiva
23 Feb.–23 April	Public hearing of the Posiva's EIA programme
24 June	Statement of the contact authority (MTI)
1999	
26 May	DiP application and EIS submitted to the MTI
21 June–20 August	Public hearing of Posiva's EIS
21 June–19 Nov.	Public hearing of the Posiva's DiP application
25 March	Decision of the Council of the State concerning the safety of final disposal
5 November	Statement of the contact authority MTI
9 November	Public gathering concerning the DiP application in Eurajoki, organized by MTI
2000	
11 January	Preliminary safety assessment by STUK
24 January	Municipal Council of Eurajoki approves the DiP application
16 and 17 February	Two appeals against the positive statement of the municipal council of Eurajoki
23 November	Capacity of spent nuclear fuel included in the DiP application reduced from 9,000 tU to 4,000 tU by Posiva. Posiva asked the Council of State to decide the final disposal of spent fuel (approx. 2,500 tU) produced in the new NPP unit at the same with TVO's application
21 December	DiP by the Council of State about the disposal capacity of 4,000 tU
2001	
7 February	Preliminary debate in Parliament
	Parliamentary expert hearing process in the Committees
16 May	Final reading in Parliament begins
18 May	DiP ratified by Parliament by 159 votes to 3
2002	
17 January	DiP by the Council of State concerning the TVO's application for a new NPP and Posiva's application for the extension of the planned disposal facility
24 May	DiP concerning the extension of disposal facility ratified by Parliament as part of TVO's application by 107 votes to 92 maximum disposal capacity increased to 6,500 tU
2004	
29 June	Posiva starts the excavation of the first phase of the Underground Rock Characterization Facility ONKALO in Olkiluoto

Index